BUILD YOUR OWN HOME
FOR LESS THAN $15,000

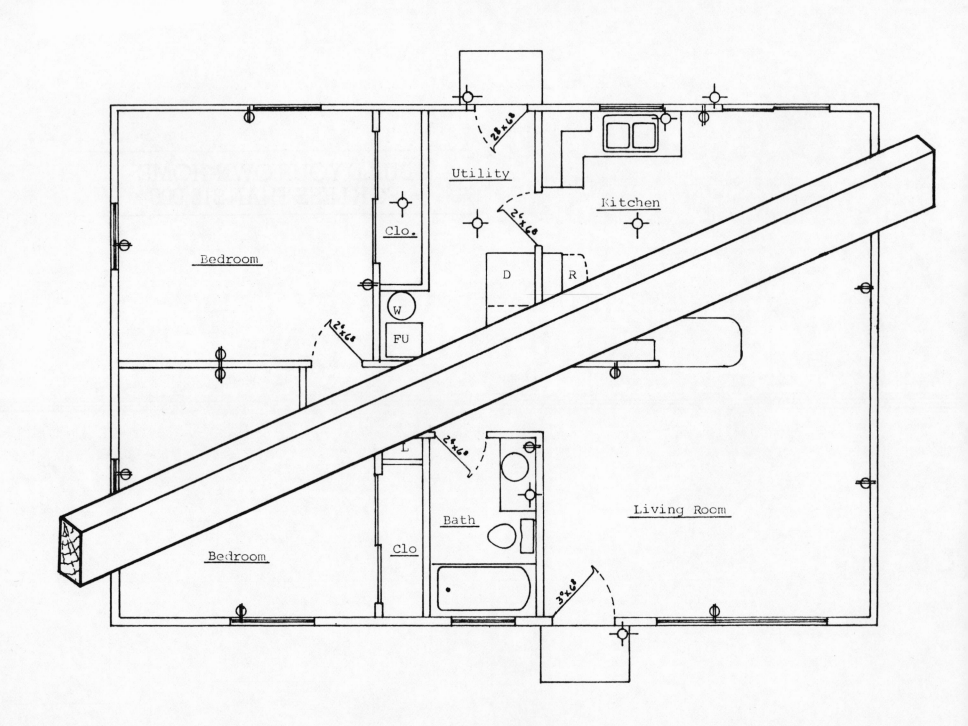

BUILD YOUR OWN HOME FOR LESS THAN $15,000

An Essential Guide to Every Phase of Building Your Own Home

PAUL BEMISH

QUILL

New York   1984

Library of Congress Catalog Card Number: 84-60483

ISBN: 0-688-02640-0

Printed in the United States of America

2 3 4 5 6 7 8 9 10

BOOK DESIGN BY BERNARD SCHLEIFER

CONTENTS

BUILD YOUR OWN HOME
FOR LESS THAN $15,000

WHY BUILD YOUR OWN HOME?

The heaviest financial burden any family bears is—and has always been—its home. A place to eat, rest, love, and be loved. For this shelter, we must, in most cases today, use the largest portion of our earnings. In many cases we pay for our housing over thirty years or we collect hundreds of monthly rent receipts as long as we live.

The moment you opened this book you showed an interest in a home—perhaps your first house or maybe a larger, better one. Whether you are having a new home built and would like to see and understand how the contractor will construct it or whether you want to build your own home, this book has been written with one goal in mind: to teach and guide you through every step or phase of building, to show how you too can have a good, comfortable home at a reasonable cost. For we all have a choice. We can have our house built by others, or we can devote a year of our spare time to building it ourselves for one third the cost leaving more money for other goals in our lives.

In this manual, you will be taught, step by step, every detail used by every trade in the building of a home. After the house is built, you may have discovered in the building industry a new profession that you like better than your present line of work. You could become a craftsman in any of these trades. You will find that this manual follows the exact sequence of steps by which a standard home is built. In plain shoptalk, I will guide you through the do's and don'ts and explain why each step is necessary. If you follow this manual carefully, you will come up with an inexpensive and beautiful home you can be proud of.

In the past thirty-five years, I have designed and built several hundred homes. But thousands of homes are needed today. The only way to produce so many homes at a reasonable cost is to teach as many people as possible to build what they need.

In the past, my wife and I spent several years with the Indian people on their reservations, helping them to build homes, a tribal building, and a large clinic. In a nearby community, a group of people needing homes joined together and used a self-help program I established to help each other build. With proper plans, material lists, and guidance, these people are constructing their own homes with no further assistance from me. Perhaps this same program could be set up in your community and work as well. The point is, if you need a home enough to be willing to work for it, you can build it economically and avoid the large mortgage payments over the next thirty years. In this manual I have taken all the mystery out of building a home.

Because of different climates, building codes in each state will be slightly different, especially with regard to the foundations. Most of those conditions are covered in this guide, but it must be your responsibility to study this manual completely, become aware through your building department of any local code variations, and work from a good set of blueprints. If, while building your new home, you find yourself unsure of your next step, just go down the street and check out another new home in progress—or contact your building department.

It has been my privilege to be a part of the birth of some of the first All Electric Homes, the Pre-Fab, Pre-Cut factory-built homes, and the factory-built Kitchen-Bath Core Units. I have also, on occasion, helped decide FHA Minimum Property Standards. All of this experience I pass on to you in this manual.

YOU CAN BUILD YOUR OWN HOME, IF YOU WANT TO!

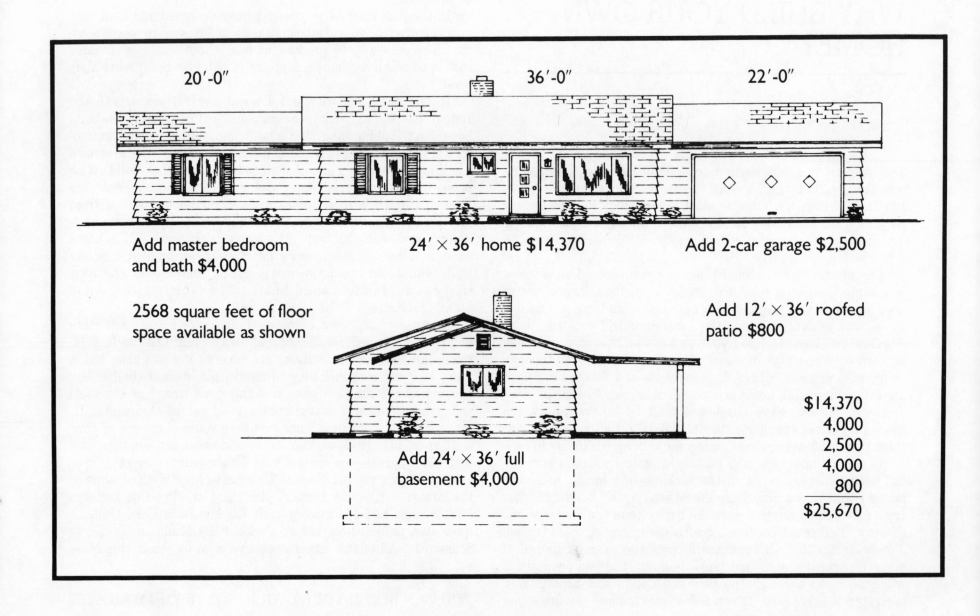

20'-0" 36'-0" 22'-0"

Add master bedroom
and bath $4,000

24' × 36' home $14,370

Add 2-car garage $2,500

2568 square feet of floor
space available as shown

Add 12' × 36' roofed
patio $800

Add 24' × 36' full
basement $4,000

$14,370
4,000
2,500
4,000
800
_____
$25,670

24′ × 36′-HOME MATERIAL COST AS SHOWN: $14,370

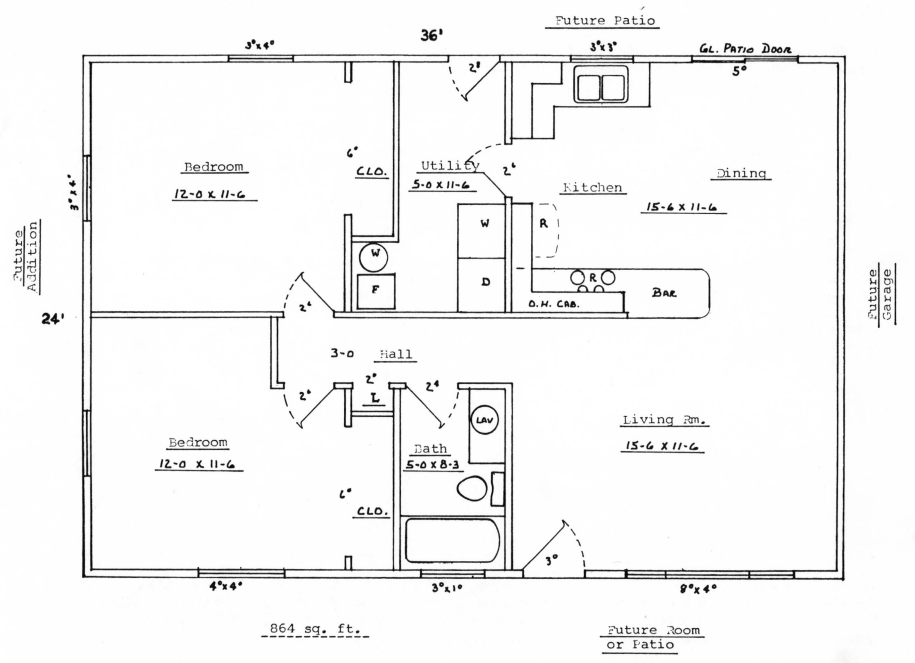

Future Patio

36′

3°x3°   GL. PATIO DOOR

5°

3°x4°

2′

Bedroom
12-0 X 11-6

6°
CLO.

Utility
5-0 X 11-6

2°

Kitchen

Dining
15-6 X 11-6

3′x4′

W   R

Future Addition

W

D   R

24′

F

2°

D.H. CAB.

Bar

3-0   Hall

2°

2°   L   2°   2°

2°

LAV

Living Rm.
15-6 X 11-6

Future Garage

Bedroom
12-0 X 11-6

Bath
5-0 X 8-3

6°
CLO.

3°

4°x4°

3°x10°

8°x4°

864 sq. ft.

Future Room
or Patio

ACTUAL COST OF MATERIALS FOR 24' × 36' HOME ILLUSTRATED

The following prices are based on the cost of materials in Southern California and should be considerably less in most parts of the country.

Concrete: slab and footings, 15 yds. at $45 yd.	$ 675.00
Concrete: forming and pouring labor	$ 864.00
Ground-plumbing materials: hot, cold, and drain	$ 375.00
Treated bottom plates, 250 lin. ft. 2″ × 4″ × 14′	$ 95.00
Double top plates, 500 lin. ft., 2″ × 4″ × 14′	$ 180.00
Headers, 58 lin. ft., 4″ × 6″; 15 lin. ft., 4″ × 8″	$ 90.00
Studs, 2″ × 4″ × 92¼″, 240 pcs.	$ 375.00
Corner braces, 1″ × 6″ × 12′	$ 40.00
2 Gable-end trusses, 4⁄12, 24 ft.	$ 90.00
17 Regular trusses, 4⁄12, 24 ft.	$ 670.00
Facia board, 2″ × 6″ × 14′, 130 lin. ft.	$ 80.00
½″ Sheathing plywood, 35 sheets	$ 360.00
15 lb. Roof felt and tar	$ 66.00
235 lb. 3 tab roof shingles	$ 350.00
½″ Celotex siding or sheathing, 4′ × 8′, 30 pcs.	$ 200.00
All windows	$ 420.00
All doors, frames, trim, locks, etc.	$ 690.00
Exterior siding, 30 4′ × 8′ sheets	$ 425.00
Wiring materials: panel, wire, switches, plugs	$ 700.00
Top-out plumbing: fixtures and vents	$ 1,000.00
4″ Wall insulation, R-11	$ 192.00
6″ Ceiling insulation, R-19	$ 260.00
½″ Drywall, 100 4′ × 8′ sheets	$ 480.00
Drywall tape and mud	$ 60.00
Texture walls and acoustical ceilings	$ 400.00
Interior paints	$ 125.00
Exterior paints	$ 70.00
Furnace, heating	$ 1,200.00
Kitchen cabinets	$ 900.00
Formica counter tops	$ 250.00
Floor coverings: carpet and floor tile	$ 1,288.00
Septic tank if not on sewer	$ 1,400.00
Total material	$14,370.00
Two-car garage	$ 2,500.00
Master bedroom & bath	$ 4,000.00
Roofed patio	$ 800.00
Basement, 24′ × 36′	$ 4,000.00
	$25,670.00

The above quantities and prices have been compiled to show that you can have a beautiful home at a fair price, provided you build it yourself. If you were to have this $25,000 package built in Southern California, it would cost approximately $90,000 on today's market. Most of us must have a mortgage on our home for many years, but the smaller the mortgage, the smaller the monthly payments. IT'S WORTH THE EFFORT.

FLOOR PLAN FOR PROPOSED FUTURE ADDITIONS

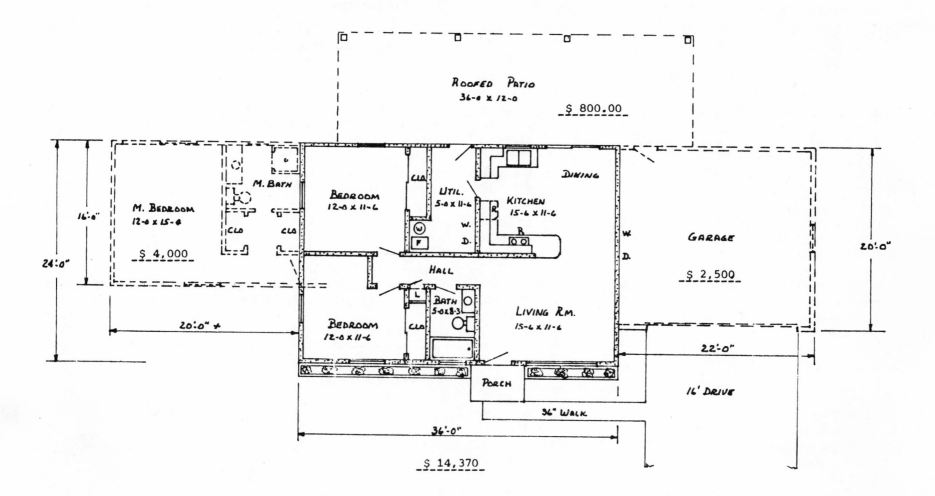

ROOFED PATIO
36-0 x 12-0

$ 800.00

M. BATH

M. BEDROOM
12-0 x 15-0

$ 4,000

BEDROOM
12-0 x 11-6

CLO

UTIL.
5-0 x 11-6

DINING

KITCHEN
15-6 x 11-6

GARAGE

$ 2,500

CLO

CLO

W

F

W.

D.

HALL

BEDROOM
12-0 x 11-6

L

BATH
5-0 x 8-3

CLO

LIVING RM.
15-6 x 11-6

PORCH

16' DRIVE

36" WALK

16'-0"

24'-0"

20'-0" ±

20'-0"

22'-0"

36'-0"

$ 14,370

You furnish the labor.

MATERIAL LIST

Master Bedroom and Bath:

Concrete: slab floor and footings, 6 yards	$ 270.00
Concrete: forming and pouring labor, average $1.00 per square foot	
Ground-plumbing: hot, cold, and drains	$ 160.00
Anchor bolts ½ × 10 (12)	$ 8.00
Bottom wall plates, 2 × 4 × 14 (100) lin. ft.	$ 33.00
Double top plates, 2 × 4 × 14 (200) lin. ft.	$ 48.00
Headers, 4 × 6 over windows and doors (20) lin. ft.	$ 24.00
Wall studs, 2 × 4 × 92¼ (100) pcs.	$ 150.00
Corner braces, 1 × 6 × 12 (48) lin. ft.	$ 9.00
(1) Gable end truss ⁴/₁₂ (16) ft.	$ 35.00
(11) Regular roof trusses (16) ft. ⁴/₁₂	$ 330.00
Facia board, 2 × 6 × 14 (64) lin. ft.	$ 32.00
Roof sheathing plywood, ½ × 4 × 8 (15) sheets	$ 150.00
Roof felt (15)# and tar, (5) rolls	$ 60.00
Roof shingles, 235# (3) tab (5) squares	$ 135.00
Celotex siding ½ × 4 × 8 (14) sheets	$ 112.00
Windows (3) Doors, frames, and trim (4)	$ 250.00
Exterior siding, 4 × 8 sheets (14) pcs.	$ 175.00
Wiring materials: wire, boxes, switches, plugs	$ 125.00
Top-out plumbing, fixtures, and vents	$ 450.00
Wall insulation 4″-R-11 (450) sq. ft.	$ 85.00
Ceiling insulation 6″ R-19 (320) sq. ft.	$ 90.00
Drywall, ½ × 4 × 8 (42) sheets	$ 180.00
Drywall tape and mud	$ 40.00
Texture walls and acoustical ceilings	$ 250.00
Interior and exterior paint	$ 75.00
Heating: additional ducts and registers	$ 150.00
Floor covering: carpet, 36 sq. yds.	$ 432.00
Total material cost	**$3,858.00**

Two-Car Garage:

Concrete: slab floor and footings, 8 yds.	$ 360.00
Concrete: forming and pouring labor, average $1.00 per square foot	
Anchor Bolts, ½ × 10 (13) on 6′-0″ centers	$ 8.50
Bottom wall plate, 2 × 4 × 14 (48) lin. ft.	$ 16.00
Double top plates, 2 × 4 × 14 (128) lin. ft.	$ 30.00
Headers, (2) 4 × 6 × 4′ (1) 4 × 14 × 16′	$ 45.00
Wall studs, 2 × 4 × 92¼″ (55) pcs.	$ 82.50
Corner bracing, 1 × 6 × 12 (3) pcs.	$ 9.00
1 Gable end roof truss (20) ft. ⁴/₁₂	$ 40.00
11 Regular roof trusses, (20) ft. ⁴/₁₂	$ 418.00
Facia board, 2 × 6 × 14′ (72) lin. ft.	$ 35.00
Roof sheathing plywood, ½ × 4 × 8 (18) sheets	$ 180.00
Roof felt 15# and tar (6) rolls	$ 70.00
Roof shingles, 235# (3) tab (6) squares	$ 162.00
Celotex siding, ½ × 4 × 8 (14) sheets	$ 112.00
Window, Service Door, and (16) ft. garage door	$ 250.00
Exterior siding, 4 × 8 sheets (12) pcs.	$ 150.00
Wiring materials, wire, boxes, switches, plugs	$ 100.00
Wall insulation 4″ R-11 (380) sq. ft.	$ 70.00
Ceiling insulation, 6″ R-19 (440) sq. ft.	$ 120.00
Drywall, ½ × 4 × 8 (28) sheets	$ 120.00
Drywall tape and mud	$ 40.00
Exterior paint	$ 27.00
Total material cost	**$2,445.00**

DESIGNING YOUR HOME

You are now ready to take the most important step in building your new home. Your choices for design, appearance, and livability can make this a happy place to live and a valuable piece of property if you should decide to sell later. As you plan, you must keep in mind that you may indeed decide to sell. Would others like your home? Would it move quickly?

This is the time to review all the clippings of ideas that you have collected from magazines and other sources. By building your own home, you can incorporate many of these ideas—for your kitchen, bath, and bedroom—into your plan at a reasonable cost. Do not get carried away, however, since these extras still cost money. Check your budget; perhaps some of these ideas can wait until later.

You should review the basic 24′ × 36′-home floor plan that we will be using throughout this manual. Consider including the four additions shown there, and you can have as nice a home as anyone on the block, for one-third the cost.

Before you start out, read through this book carefully. Then you will be ready to draw your own floor plan and room arrangement. Refer to the drawings on page 13 as a guide. You will need ¼″ graph paper, a pencil, a 12″ scale or ruler, and a triangle. Make your drawing to a ¼″ scale that equals one foot. In establishing the outside dimensions of your home, remember that all lumber comes in even foot lengths: 8′, 10′, 12′, 14′, and so on. Do not come up with a 27′ depth of house. This would require cutting off and wasting one foot of every floor joist, ceiling joist, etc. Make your home 28′ deep.

To help establish room sizes and details, consider each room separately; now is the time to remember every detail you want to include. Refer to the plot plan again. Do not come up with a home too long for the width of your lot.

KITCHEN

There are two basic plans for designing a kitchen. One is a working kitchen, with no table area. You might include a breakfast bar for snacking, but all regular meals would be served in a dining area in another room. The other plan is to design a kitchen large enough for a table and chairs. This could eliminate a formal dining room. Decide now which plan best fits your life-style.

Next, draw in your cabinets to scale, in the desired locations. Allow 24″ from the wall to the face of the cabinets. Refer to the section "Kitchen Cabinet Details" to help you decide what kind of cabinets you'll use.

Next, locate the double sink, range, and refrigerator—24″ × 36″ space for both appliance units. Over the sink, call for a 36″ × 36″ window. Try to come up with a pantry for food storage. It could be recessed into a bedroom or utility closet. If the table will be in the kitchen, you should have a large window down to the table height or perhaps a sliding glass patio door. You should also include a door from the kitchen to the utility room; then plot the outside door. DREAM ON—it's worth it. The kitchen is a very important, much used room; it deserves much thought.

UTILITY ROOM

If you do not plan on a basement or garage at this time, a utility room is a must. In this room or in the garage you must install the furnace, water heater, washer and dryer, leaving enough room for a storage area. If part of this equipment can be installed in the garage, you can reduce the size of the utility room.

LIVING ROOM

You will spend more time in this room than in any other room during your waking hours each day. This room should not be less than 12' wide in any portion. Most living rooms are close to 15' wide. NOTE: Carpet comes in 12- and 15-foot widths.

If at all possible, try to design your living room so you can avoid going through that room every time you go to the bathroom or bedroom. Many times this is impossible unless you have a long rambling type of home in mind.

You may desire a fireplace in the living room. Now is the time to plan for it. You can purchase a factory-built metal unit to build in during framing for less than $500. This fireplace will help heat your home and will certainly increase the value of the house.

For an average 15' × 18' living room, you should install a picture-window unit 8' wide by 4' high for light and ventilation. Your front door should be 3' wide, 6' 8" high, the standard size. The living-room floor is usually inlaid carpet with a good-quality pad underneath.

BEDROOMS

Your needs must determine the number of bedrooms in your new home. No bedroom should be less than 10' × 10'. A second bedroom should not be less than 12' × 12' for room arrangement or livability. If you plan a third or master bedroom at this time, give this area a lot of thought. NOTE: Every bedroom must have at least one window large enough to climb through in case of fire. The bottom of this window should not be more than 44" above the floor. A good standard-size window commonly used in bedrooms is 48" wide and 36" high. You may want to use a 48"-high window for better exterior appearance and ventilation. When a bedroom has two outside walls, locate a window in both outside walls. If a bedroom has only one outside wall—and thus one window—use a wider window, such as a 60"-wide, 36"-high unit.

Master Bedroom: Design this room so it is large enough for at least three different furniture arrangements. Plan your wall space to accommodate various pieces of furniture and plan a pleasing window arrangement that allows for as much cross ventilation as possible. This room must also have ample closet storage space, preferably a closet with a door. If you choose to have a master bedroom, seriously consider a private master bath.

Bedroom Closets: Every bedroom must have a closet. You should allow for a depth of 24" for all closets. Make the closets as long as possible. Most people never have enough storage space. If you come up with a closet only 48" long, you should install a 24"- or 30"-wide, swing-out door. If the closet is considerably longer, you could install a by-pass or by-fold door. These types of doors come in 4'-, 6'-, or 8'-wide units. If you install your clothes pole 60" above the floor, you will have room for two 12"-deep shelves, for the full length of the closet. You can install these shelves by using special metal shelf brackets that will also support the clothes poles. Perhaps you will want ceiling lights in your master-bedroom closets. Think about two separate closets in the master bedroom.

BATHROOMS

Your main bathroom will serve your family and guests. Not too small, it should include a tub/shower combination with curtain or sliding tub door. The one-piece fiberglass tub/shower combinations are very popular in most areas, BUT this tub must be set in place during framing and standing of walls. Because of its size, it cannot be set in place later, unless you remove some studs from the outside wall.

You should allow for installation of a lavatory and cabinet. This unit is usually constructed of wood and has a marble, tile, or Formica top surface. Next you should plan to include a stool, a medicine cabinet or a mirror in the wall, and lighting. Call your local stores for measurements. Your master bathroom will be set up basically the same way. You may prefer just a tub or a shower rather than a combination.

HALLWAY

Design your hallway to be as short as possible but never less than 36″ wide. The only purpose of a hallway is to avoid going through one room to get to another. In this hallway, include a linen closet 18″ deep with five shelves and an 18″-wide door. Recess this closet into the end of a bedroom closet near the bathroom.

GARAGE

A new home should include a two-car garage. It will cost only a little more at this time, yet it will increase the value of your home so much. Also, you may want to install your furnace, water heater, and laundry facilities in this area. You should still have space for your auto and extra storage.

Your garage should not be less than 22′ deep and 22′–24′ wide. This will give you plenty of room. The wall between the living area and attached garage should be a firewall-type construction. You should install a 32″-wide solid-core door between the living area and garage. You should also have a 32″-wide service door from the garage to the outside. A 48″×36″-high window in one wall will allow for natural light. For a two-car garage, install a standard-size overhead door— 7′ high by 15′ wide and made of wood or metal.

FLOOR PLANS AND BLUEPRINTS

You now have a floor plan laid out for your desired home. You could take this drawing to a local designer and tell him, "This is it." However, if you follow the drawing in this manual you can prepare your own drawings with the proper guidance from your local building department. This is a part of their work—to guide you in accordance with the code.

First, referring to your floor-plan drawing, you have to determine the outside dimensions of your home. These figures, along with your side-yard dimensions, tell you how wide a lot you must have to build your home on.

If you decide to make your own drawings, be sure to use regular paper that will blueprint. Purchase a vellum-type tracing paper from your local stationery store. Drawings should be 18″×24″ or 24″×36″ for blueprinting. You will need at least three sets of prints for permits and for building your home.

YOUR BLUEPRINTS SHOULD COVER THE FOLLOWING:

1. Plot plan drawing and legal description.
2. Floor plan showing all dimensions, electrical layout, window and door sizes, headers, and room sizes.
3. Foundation layout showing all dimensions. Material sizes and descriptions should be shown on drawings.
4. All exterior elevations of home, showing all exterior details, including type of windows, siding, and roof.
5. Wall-section details showing every detail from bottom of footings to top of roof.
6. Any special construction details required in building this home, garage, or patio.
7. Complete kitchen-cabinet layout, ready for bids.
8. A complete, detailed material list.

NOTE: If you would like to build the 24′×36′ home illustrated in this manual and don't want to prepare your own plans and blueprints, you can order them from:

Home Planning Service
P.O. Box 818
Indio, California 92202

619 555 1212

Include $20 per set, check or money order, for the prints and mailing. When ordering blueprints, advise which additions you will add to the basic 24′×36′ home. Also include which type of siding you will use—panel, lap, brick, or stucco.

SELECTING YOUR LAND

Selecting the location on which you will build your home is very important. You will no doubt live in this home and neighborhood for many years. Possibly you have already decided that you would like to live in a certain area. You should have many questions. Is this location suitable to raise a family? Is there an abundance of children there now? How far must you drive to work, shopping area, schools, church? How far away are fire and police protection? When you have answered all your questions, you will know if you have chosen the right neighborhood.

A choice plot of land should be slightly higher than the street. It should not show any signs of flooding or drainage washes nearby. Are the surrounding homes well kept? Do they reflect pride of ownership?

Before buying this property, the following should be checked out. Is a public water supply available to this property? Must you drill your own well? Is a sanitary sewer available to tie into, or must you install your own septic tank? Are there any easements across this land that would restrict your home size? Your real-estate agent can give you most of these answers, and your building department can furnish the rest. Your side-yard setback dimension is also very important. Again, your building department will tell you; it can run from six feet on up, on both sides of house. This means, for example, that you cannot build a fifty-foot-long home on a fifty-foot-wide lot.

On the plot-plan layout, fill in the dimensions of your home and lot size. Check out your setback lines side and front; fill these in and this will tell you if your home will fit the lot you've chosen.

Let your real-estate agent handle the paperwork for the property. In some areas you will have a deed abstract, title search, and insurance. It is worth the money to be totally protected.

Legal Description:

Lot No._____ Street No._____

Section_____ Range_____ Township_____

County_____ State_____

PLOT PLAN OF YOUR PROPERTY

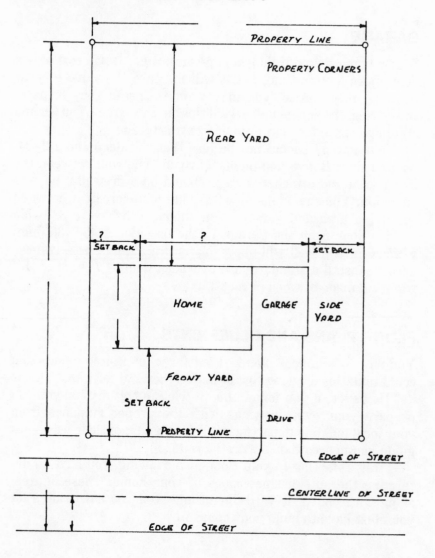

CONCRETE FLOOR SLAB HOME

MATERIAL LIST

The following list of materials covers all items required from the bottom of the footings to the floor.

Description (House)

		Quantities	Cost
Concrete footings	cu.yd.		
*Visqueen vapor barrier	sq. ft.		
Concrete floor slab	cu.yd.		
Wall anchor bolts, ½″ × 10″			
*Wire mesh, reinforcing steel	sq. ft.		
*Reinforcing-steel bars, ½″	lin. ft.		
Porch or patio slab	cu. yd.		
Concrete sidewalks	cu. yd.		
*2″ sand under slab	cu. yd.		
Steel anchors/porch posts			
*Styrofoam insulation	sq. ft.		

Labor Cost

Digging and forming footings	
Installing Visqueen and reinforcing bars	
Installing wire mesh and sand	
Forming porches and walks	
Pouring and finishing concrete	
Installing ground plumbing	
Forming, pouring, and finishing driveway	

*These items may not be required in your area.

Description (Garage)

		Quantities	Cost
Garage footings	cu. yd.		
*Visqueen vapor barrier	sq. ft.		
Concrete floor slab	cu. yd.		
Wall anchor bolts, ½″ × 10″			
*Wire mesh, reinforcing steel	sq. ft.		
*Reinforcing steel bars, ½″	lin. ft.		
*2″ sand under slab	cu. yd.		
*Concrete driveway	cu. yd.		
*Expansion joints	lin. ft.		

Refer to plumbing chapter for necessary plumbing materials that must be installed before concrete is poured.

Refer to footing section drawings covering proper details for your home. This detail is determined by type of exterior of your specific home. Brick veneer, lap siding, stucco, etc.

MATERIAL LIST: COST BREAKDOWN

You are now ready to learn a new profession. In the following pages you will learn to figure the material list not only for your new home but for others as well. With this knowledge, you will be able to apply for a position with your local lumber company as a material estimator. You will be prepared to figure their bids on a package home, including all materials that a lumber company normally supplies.

If you devote many hours of study to this chapter, you will become acquainted with every detail involved in building a home.

Each of the following paragraphs provides a step-by-step description of every item as it is used during the construction of a normal family home. Each item in the following paragraphs will be listed on the various material lists that follow this chapter.

As you study each paragraph, refer to the proper wall-section drawing. This will enable you to identify and locate the proper use of each item as required during construction.

We will start off with a basement home with a wood floor. If your home will not include a basement, pass over these in-

*These items may not be required in your area.

structions and proceed to the proper description of your home—wood floor with crawl space or concrete slab floor.

Excavate Basement

At this time, you are ready to start saving money. You should obtain at least two bids from local contractors for digging this hole. Do not attempt this project by hand. To figure the cubic yards of earth to be removed, use the following formula. If the outside dimensions of your home will be 20′ × 40′ and 8′ deep, dig the hole 2′ larger than the house on all sides. This extra size is required to allow room for waterproofing outside basement walls. You now need a hole in the ground 24′ × 44′ × 8′ deep, 24′ × 44′ = 1,056 sq. ft. × 8′ deep = 8,448 cu. ft. There are 27 cu. ft. in a cubic yard; 8,448 cu. ft. divided by 27 cu. ft. = 312 cu. yds. of earth to be removed.

Reinforcing Bars

In some areas, re-bars are not required in the footings. Check your plans to see if they have been called for. If in doubt, figure in two ½″ bars completely around the footings. You will need (2 × 120′) 240 ft. of ½″ re-bar.

Concrete Footings

Footings will project outside of basement walls about 6″. (See wall-section drawing.) For a 20′ × 40′ home you will probably use an 18″-wide by 12″-deep footing. Each running foot of this size footing will use 1⅓ cu. ft. of concrete. (The distance around your 20′ × 40′ home is 120′, so 120 × 1⅓ = 160 cu. ft. of concrete; 160 divided by 27 = 6 cu. yds.) You will need 6 cu. yds. of concrete for the footings.

¾″ Rock for Footing Drain

Figure that you will need a truckload of this rock. Any rock left over can be used in a septic system. In some areas, this rock and 4″ footing drains are not required. Check with your building department; it's a good investment, though, and cannot be done later. Spread the rock out 6″ deep around the outside of the footings and about 24″ wide. Before installing the 4″ tile, waterproof the outside of the basement walls. (See paragraph on Mastic Mop and Visqueen Vapor Barrier.)

4″-Tile Footing Drain

You now have your basement footings poured. At this time you will install the 4″ red-clay tile around the outside edge of the footings. These tiles are 12″ long and should lie end to end completely around the basement footings. Use one clay elbow at each corner of the footings. You will need one clay "Y" to go under the footing and into the sump pit. Figure in your pit and cellar drainer at this time. Hand-dig a trench about 24″ wide and 6″ deep below the top of the footings around the outside. Fill with ¾″ rock. Lay tiles tightly end to end with the centerline of the tiles even with the top of the footings. Cover each tile joint with a 6″ × 12″ strip of 15 lb. black-felt building paper. You are now ready to place another 6″ of rock over the tile.

Vertical Steel Reinforcing Bars

Some areas may not require this steel. If you do need them, place these ⅜″ steel bars full height vertically every 32″, completely around basement walls. Your 20′ × 40′ home has 120′ of basement walls which will require 45 bars 8′ long. (See paragraph on grouting concrete for next step.)

Concrete Wall Blocks

If your home will be brick or stone veneer, you will no doubt use blocks that are 12″ wide, 8″ high, and 16″ long. With this size block, some areas may not require the ⅜″ vertical steel bars. Verify block size with your local building codes.

Since it is 120′ around the house and the blocks are 16″ long, one row of blocks around the house will require 90 blocks. A normal basement ceiling height is 7′10″ finished. This would call for (12 rows of blocks, 90 per row) 1,080 blocks. (See proper wall section.) The top row will be smaller if you do use brick or stone veneer.

Mortar and Sand

When shopping for quantity and prices for blocks, ask your local block supplier to include the cost for mortar and sand to lay blocks and for grouting the walls.

Grouting Concrete

You should purchase your grouting material along with the mortar and blocks, all in one load. Your grouting material is made up of cement and sand and should be mixed thin to pour into the cavity of the blocks around the ⅜″ vertical steel bars. Fill the cores with grouting after each 3 rows of blocks are laid.

Basement Windows

There are various types of steel- and aluminum-frame basement windows to choose from. Each wall of the basement should have at least two windows, except the wall at the front of the house. A standard window is usually 36″ wide and 24″ high. These windows are installed as the block wall is laid up. Order them along with the blocks, mortar, grouting, and sand.

With each basement window, you will need a metal window well. This is a half circle of metal, about 24″ high, which fastens over the window on the outside basement wall. By using a window well, you can finish-grade your yard almost up to the top of the window. (See detail drawings.) After the window wells are installed and the outsides of the walls are backfilled, cover the floor of the well with 2″ of ¾″ rock to allow for drainage. DO NOT BACKFILL AT THIS TIME. Walls could push in at this stage.

Masonry Wall Bolts

These bolts are ½″ in diameter and 10″ long with a square bend on one end and a nut and washer on the other end. These bolts are set in the grouting in the cores of the top row of blocks. Allow 2½″ of the threaded end to project up above the top of the block walls. These bolts secure the wall plate and walls to the basement block walls.

Figure one bolt to be installed every 6′ around the entire top of the basement walls. Also, figure two bolts for each corner, one 12″ in each direction from the corner. These bolts will not always sit directly on the center of a block width. (Study this detail drawing closely.)

Mastic Mop and Visqueen Vapor Barrier

You are now ready to waterproof your outside basement walls. A 5-gallon bucket of waterproofing mastic should cover about 200 square feet of wall. With a wall 7′ high and 120′ around, you have 840 square feet of wall to cover. Four buckets should take care of it. Figure on one box of Visqueen. This is applied like wallpaper over the wet mastic. The basement is now sealed from outside moisture.

Termite Shields

This item may be required in your area. (Check your plans.) If termite shields are required, your local sheet-metal shop can fabricate them for you in accordance with the code. The material used is 26-gauge galvanized sheet metal. When installing these, it will be necessary for you to mark and cut holes in the metal, allowing the shield to fit down over the masonry bolts. Lap the joints and tar them over.

Visqueen Under Basement Floor

In some areas, this Visqueen is required. Check your area. If Visqueen is required, your 20′ × 40′ home will need 800 square feet.

After the level of the basement floor is determined (3½″ above the top of the footings) remove 2″ of earth down from the top of the footings.

Next, cover the entire floor with Visqueen and cover that with 2″ of sand. This will bring the floor back up flush with the top of the footings. This sand surface must be smooth and level to receive a 3½″-thick concrete floor supported by the footings around the outside of the basement. At this time you will dig holes for the 18″ × 18″ × 12″-deep piers every 8′ the full length of the basement under every jack post. (See detail drawings.)

Basement Concrete Floor

To figure the cubic yards of concrete required to pour the basement floor of your 20′ × 40′ home, follow this formula:

For the required 800 square feet of floor, figure 3½″- to 4″-thick concrete with shrinkage included. One cubic foot of concrete will pour 3 square feet of the floor. A cubic yard equals 27 cubic feet; $3 \times 27 = 81$ sq. ft. of floor poured per yard of concrete. Then divide the 800 square feet of floor needed by 81. This means you will need 10 cubic yards of concrete to pour this floor; 10 times the local cost per yard gives the cost for the basement floor.

6″ × 8″ Built-Up Girder or 6″ Steel I Beam

This beam or girder will be 40′ long. It will be installed down through the center of the home, 10 feet back from the front wall. This beam will rest on the eleventh row of blocks. The block wall at both ends of this beam will be grouted full height. The purpose of this beam will be to carry the full load of the floor, walls, and roof down through the center of the house.

The top of this beam or girder must be set 1½″ above the top of the top row of blocks. This can be done by installing a 2 × 6 flat on top of the beam or girder. This beam will then match up with the top surface of the 2 × 6 wall plates, ensuring a level floor joist. (See detail drawings covering this point.)

If a 40′-long steel I beam is used, this beam will weigh 11.5 lbs. per foot. (Verify with local codes.) If a 6 × 8″ beam is used, the following 2 × 8 lumber is required: three 2 × 8 × 8′ long and six 2 × 8 × 16′ long. This would make three 2 × 8's 40′ long. These nine 2 × 8 boards will be bolted together with ½″ diameter bolts every 36″ along the full length. All joints in this built-up beam must be staggered and on 8′ centers. This will allow all joints to be over a jack post. (See drawing.)

Adjustable Jack Posts

When setting the built-up beam or I beam in place on each end of the block walls, the adjustable jack posts should be set in place every 8′. The top flange of the jack post should be anchored to the bottom of the beam with ⅜″ × 2″ lag screws.

The bottom flange of the jack post should be anchored into concrete with same-size lag screws and shields. With a level on top of the beam, adjust each jack post until the beam is level all along its length. Four jack posts will be required for a 40′ span. (Review drawings for details.) NOTE: If the steel beam is used, it will be necessary to have this beam delivered and set in place by a crane. As an alternative, you could install two 20′ beams, then weld the joint over a jack post. If a wood beam is used, this beam can be built in place, using three 2 × 8's bolted together, starting at one end wall and out to the first jack post. As the beam is built up or installed, brace it from the front and back basement walls, using 2 × 4's on top of the beam and the block walls. When bracing, pull a chalk line the full length of the beam to keep the beam straight. Put your level on each jack post to see that each one is plumb.

Bottom Plates

The bottom plates will usually be of 2 × 6 × 12′ or 14′ lumber. These plates lie on top of the block walls completely around the basement or outside of the house.

For your 20′ × 40′ home, you will need 120 running feet of 2 × 6's for this plate, end to end around the house. This would be ten 12′-long 2 × 6's.

Next, lay these 2 × 6's on top of the termite shield, with the edge of the plate tight against the masonry bolts. Cut the 2 × 6's to the proper length to lie flat and tight around the wall. Mark the exact location of each bolt on the plate, then measure at every bolt the distance from the outside edge of the block wall to the center of each bolt. Mark this dimension also on the plate.

This locates the exact drilling point for each bolt. Drill a ¾″ hole through the plate for each bolt, then place each board down over the bolts. DO NOT INSTALL FLAT WASHER AND NUT AT THIS TIME.

Box End Plates

You now have the 2 × 6 bottom plates cut, drilled, and set down over the anchor bolts in the top of the block basement walls. Since most floor joists run from the front to the back of a home, you will need box end plates only along the full length of the front and back.

Your model home is 40′ long. As a result, you need 80 running feet of 2 × 8 box end plate. This would be seven 12′-long 2 × 8's with extra. These box end plates are always the same size as the floor joist. (Refer to the drawing showing box end plate installation.)

You will note that the box end plate stands on edge on top of the outside edge of the 2 × 6 bottom plate. Nail the bottom plate to the bottom edge of the box end plate, using a 16p nail every foot. This will form an L-shaped assembly ready to set down over the wall anchor bolts. Your floor joist will rest on top of the bottom plate and butt against the box end plate. Drive three 16p nails through the face of the box end plate and into the end of each floor joist. Toenail both sides of the floor joist into the bottom plate.

Floor Joists

To figure the number of floor joists required in your 40'-long home, start at one end of the home, laying the first joist flush with the outside block wall and on top of the 2×6 bottom plate.

Next, measure 16″ from the outside edge of the first joist to find the center of the second joist, and 16″ thereafter for each joist. This will be every 16″ for the full length of the house and must be exact. You will need 31 floor joists on a 16″ center across the front of the home. You will also need 31 joists across the back half of the home. All walls running in the same direction as the floor joists must have two extra joists under each wall, no matter where they come. This is to support the wall and roof loads.

Add these to the 62 joists already needed.

Next, add 3 additional joists for double joisting around the stairway opening. (Refer to stairway details.)

Set all the floor joists in place, one on the bottom plate against the box end plate and the other end on top of the beam down through the center of the basement. With the outside end of each joist nailed in place to the box end plate, locate and nail the other end of the floor joist down through the center of the house, directly over the beam and jack posts.

Again, these joists must be set exactly on 16″ centers. NOTE: Plywood subfloor comes in 96″-long sheets. The end of every sheet of plywood must be nailed to the top edge of the floor joist. You must keep the floor joists 16″ center to center and full length. Subfloor plywood sheets must be staggered so that no two sheets end together on the same joist next to each other. (See drawings covering floor joist and subfloor installation.)

Down through the center of the floor, along the side of the beam, nail in blocking. If the floor joists are 2×8's, cut pieces of 2×8 $14\frac{1}{2}$″ long to nail on edge between all joists down through the center of the house near the beam. When each $14\frac{1}{2}$″ block is nailed in place between the joists, then nail the lap of the front and back floor joists together. Toenail the joist to the top of the main beam over the jack posts. Stagger these $14\frac{1}{2}$″ blocks $1\frac{1}{2}$″ back and forth to allow for nailing into both ends of every block, three 16p nails at each end.

Then nail $14\frac{1}{2}$″ blocks, staggered in the same way, halfway along the length of floor joists, both front and back of the main center beam. This will make 3 rows of blocking the full length of the house between all the floor joists. Figure ten 12'-long 2×8's for blocking.

Subfloor Plywood

In your model $20' \times 40'$ home, you have 800 square feet of floor to cover. Subfloor should always run at right angles to the floor joists. EXAMPLE: Your floor joists run from front to back of your home. Your $4' \times 8'$ sheets of plywood will run 40' the long way. Use $\frac{1}{2}$″ $4' \times 8'$ sheathing-grade plywood for the subfloor.

Since you have a 40' span to cover, 5 sheets 8' long will do it. You have 20' front to back to cover, which will take 5 sheets. Results: $5 \times 5 = 25$ sheets needed to cover the entire subfloor. Figure on 2 extra sheets since you must stagger the joints of all the sheets.

Nailing of Plywood

Use twelve 8p nails at both ends of every sheet, and 8 nails on all other joists. Purchase 50-lb. boxes of 8p and 16p nails to start. They are much cheaper and you will need every one of them.

Basement Stairs

First review the basement-stairs drawing. This is one of many ways to build these stairs. This method uses two 2×10 boards 14' long for stringers. The 33″-long treads for 36″-wide stairs will also be 2×10 boards. You will also need two 4×4 posts 5' long for supports underneath. When building stairs, first make a layout as shown on the drawings and study it carefully.

LEARNING TO USE THE TRANSIT

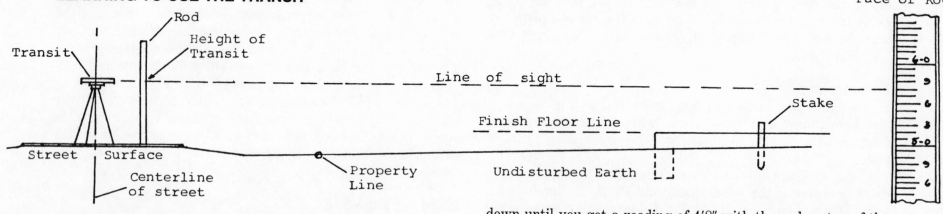

Setting of Finish Floor Level Above the Street

You are ready to establish the finish floor level above the street or road. This is a very critical step and cannot be changed later. Due to the cost of a transit, it would be wise to rent this instrument locally.

You must now learn how to read a transit. This tool is a very delicate instrument and must be handled with care. Attach the telescope head to the tripod legs, then carry the unit out in front of your home and stand it up on the approximate centerline of the street or road. Directly below the telescope, you will find four adjusting thumb screws. After turning the scope to look down the road, adjust two of these thumb screws to level on top of the transit.

Next, turn the scope toward your home. Again, level the scope in this direction with the two remaining thumb screws. This unit must be leveled perfectly in all directions.

Next, extend the rod to full length and tighten the thumb screws. Stand the rod at the front of the scope as shown above and take a reading. Example: The reading above is 5′8″ line of sight. Your reading, "Height of Transit," is very important and will be used over and over again.

Your next step is to drive a wooden stake into the ground at the approximate center of the location of your home. Have someone hold the rod on top of this stake and you can sight through the scope for a reading. If the code in your area requires your floor to be 12″ above the street, drive the stake

down until you get a reading of 4′8″ with the rod on top of the stake. This places the top of the stake 12″ above the street.

If you should get a reading of 3′8″ with the rod standing on the ground near the stake, this would tell you that your site is 24″ above the street. There is no limit on how high the finish floor can be above the street. You must decide how your home will look at this height (or whatever height you choose) above the street.

You may want to bulldoze this area down some to come out with a level pad to build on. If your land is too low to meet your code, it will be necessary to fill in your pad area to raise your floor line to obtain the correct height above the street. In most cases, if extra fill is needed, it can be taken from other spots on your land.

You have just built your pad for a concrete slab floor. This pad must be watered throughout several times to come up with a solid base to support your concrete slab.

If your home will have a wood floor with crawl space or basement, it will still be necessary to establish a floor level as explained above. With a crawl space, you will have a sizable amount of earth to remove. You can use this earth around the outside of your home to raise the elevation of the yard. A basement will supply even more earth to be spread.

NOTE: When you use a transit, if your rod reading is less than the height of the transit reading, your land is higher than the street. If the rod reading is more than the height of the transit reading, your land is lower than the street.

FOUNDATION LAYOUT

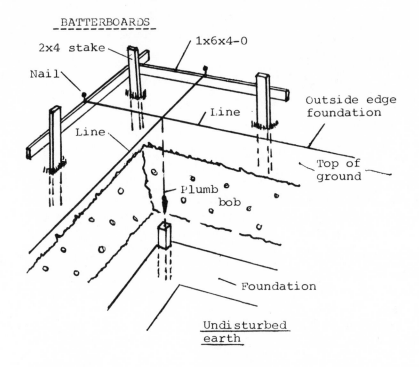

BATTERBOARDS

2x4 stake

Nail

1x6x4-0

Line

Line

Outside edge foundation

Top of ground

Plumb bob

Foundation

Undisturbed earth

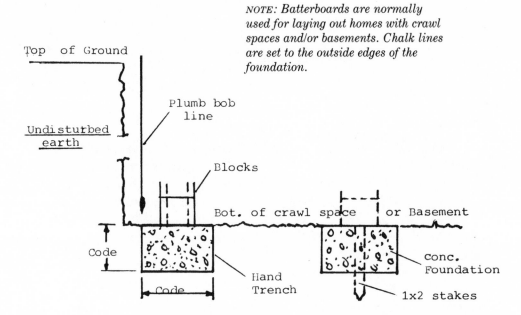

NOTE: Batterboards are normally used for laying out homes with crawl spaces and/or basements. Chalk lines are set to the outside edges of the foundation.

Top of Ground

Undisturbed earth

Plumb bob line

Blocks

Bot. of crawl space or Basement

Code

Code

Hand Trench

conc. Foundation

1x2 stakes

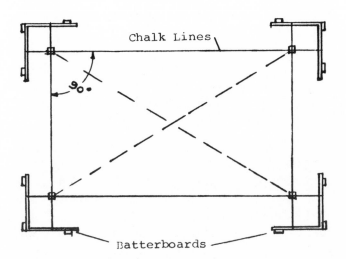

Chalk Lines

90°

Batterboards

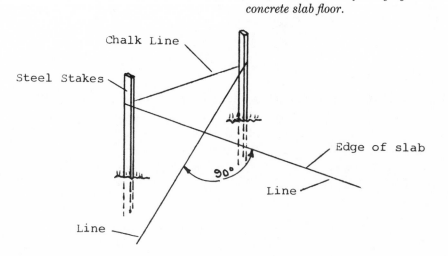

This method is used for laying out a concrete slab floor.

Chalk Line

Steel Stakes

Edge of slab

90°

Line

Line

CONCRETE FINISHING TOOLS

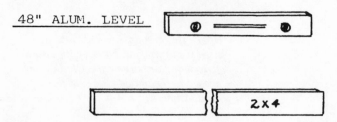

48" ALUM. LEVEL

2 X 4

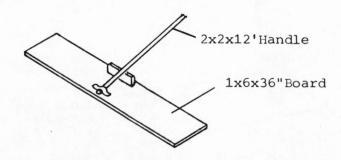

2x2x12'Handle

1x6x36"Board

STRIKER BOARD: This will be the first tool used as the forms are filled with concrete. Select a good straight 2×4, about 24″ longer than the width of the forms, as your tool. With a man on each end of this board, place the board on edge on top of the forms and with a slow seesawing motion pull the board slowly toward you. The concrete will now level off even with the top of the forms.

BULL FLOAT: The rush is over; you have the slab laid by as they say. You are ready to use the bull float over the entire slab. Watch your temper; it takes practice to use a bull float. Try this: As you push the float out, lower the handle; as you pull the float back, raise the handle. How about that?

Rental

JOINT TROWEL: Used for joints in driveways and sidewalk construction. We will cover this tool later with driveway and sidewalk forming and pouring.

HAND TAMP: After an area has been partly filled with concrete, start a man on the hand tamp. Use the tamp at once after the striker board. With an up-and-down motion, use the tamp to compact the rocks and at the same time raise the sand, cement, and water to the top. This is a must; tamp the entire area of new concrete.

WOOD FLOAT: You will notice that the water begins to disappear from the top of the slab; it is then time to use your wooden hand float. You will need two pieces of plywood about 24″ square for kneeboards. Use these boards to support your weight on the fresh concrete while using your wood float in a half-circle motion. This, again, brings the cement to the top for a smooth rich finish.

STEEL TROWEL: Your new slab is now curing out; it is time to use your steel trowel. Again, with your kneeboards under you, go over the complete slab with the steel trowel in a half-circle arc, back and forth. If you desire a very smooth slab, it may be necessary to use a steel trowel a second time. If a rough surface or a broom finish is desired on porch, patio, walks, or driveway, lightly drag a stiff-bristle pushbroom in a straight line over the slab. Then use the edging trowel.

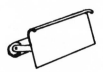

EDGING TROWEL: This trowel is normally used only on outside slabs, porch, walks, and driveway. When this trowel is used around all outside edges of slabs, square edges become rounded off to prevent chipping and ragged edges.

MORTAR TROWEL: Used for laying blocks, brick, and stone; we will cover this tool later when we discuss laying blocks.

FOUNDATIONS

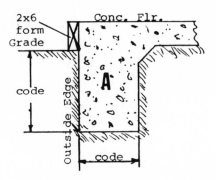

Type A: This type of foundation is commonly used in warm climates. It is usually 12″ wide and 12″ deep below finish grade, with 6″ from finish grade to finish floor. A 2×6 form is first set around the outside walls of the home, followed by a 12″-wide hand-dug trench. If brick veneer is used on outside walls, it will be necessary to add 4″ to all outside walls to rest the brick on top of the slab.

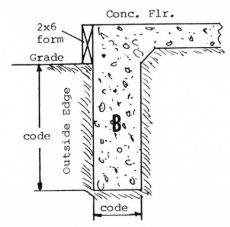

Type B: This type of foundation is used in cold climates and varies in depth according to the frostline in your area. This type is usually 12″ wide and as deep as your code requires. In most codes, you must allow 8″ from finish floor to finish grade. If brick veneer is used on outside walls, it will be necessary to add 4″ to all outside walls to rest the brick on top of the slab.

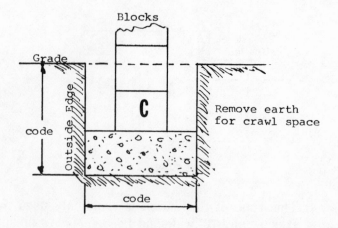

Type C: This type of foundation is used for homes with wood floor and crawl space. This type of footing requires no forms because the trench will serve as a form. Batterboards and chalk lines will be used to square up the footings. (See instructions on batterboards.) If brick veneer is used, you must add to the outside edge of the footings. **Example:** If outside to outside of frame wall is 40′ outside to outside of block, supporting walls would be 40′8″, that is, 4″ added all around to rest the brick on. From the outside edge of the block to the outside edge of the footings is normally 5″ around all outside walls; this would add another 10″ to the 40′8″; this would make outside to outside of the footings 41′6″ overall.

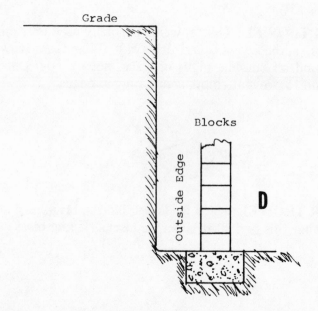

Type D: Basement footings will have same dimensional stackup as Type C. Use batterboards.

MASTER FOUNDATION PLAN FOR CONCRETE SLAB FLOOR

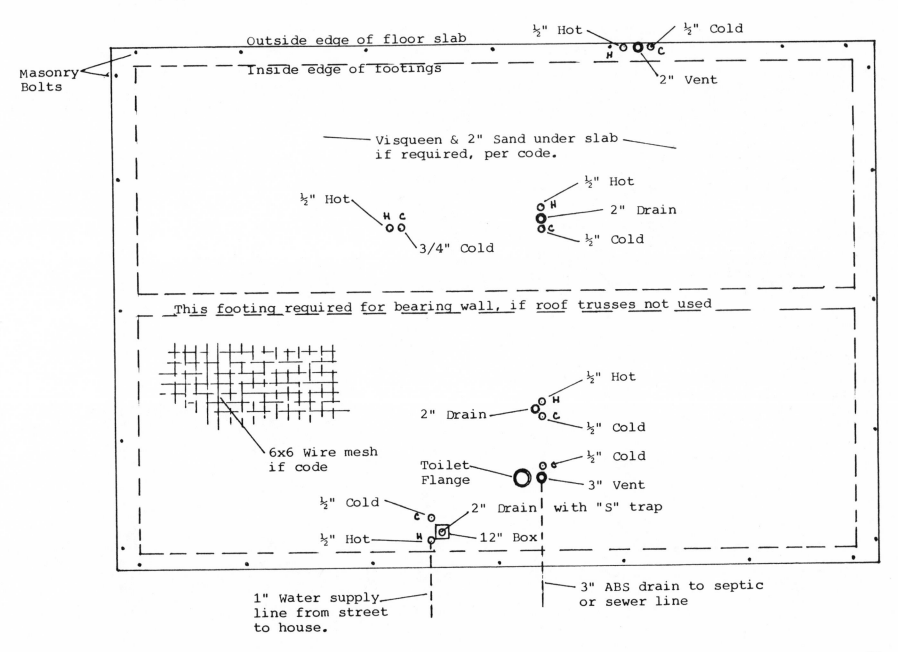

Outside edge of floor slab

½" Hot   ½" Cold

H   C

2" Vent

Inside edge of footings

Masonry
Bolts

Visqueen & 2" Sand under slab
if required, per code.

½" Hot

½" Hot   ½" Hot

H   H   2" Drain

H   C   C

½" Cold

3/4" Cold

This footing required for bearing wall, if roof trusses not used

½" Hot

H

2" Drain

C

½" Cold

6x6 Wire mesh
if code

½" Cold

Toilet
Flange

3" Vent

½" Cold

C   O   2" Drain   with "S" trap

½" Hot   H   12" Box

1" Water supply
line from street
to house.

3" ABS drain to septic
or sewer line

CONCRETE SLAB FLOOR, WARM CLIMATE

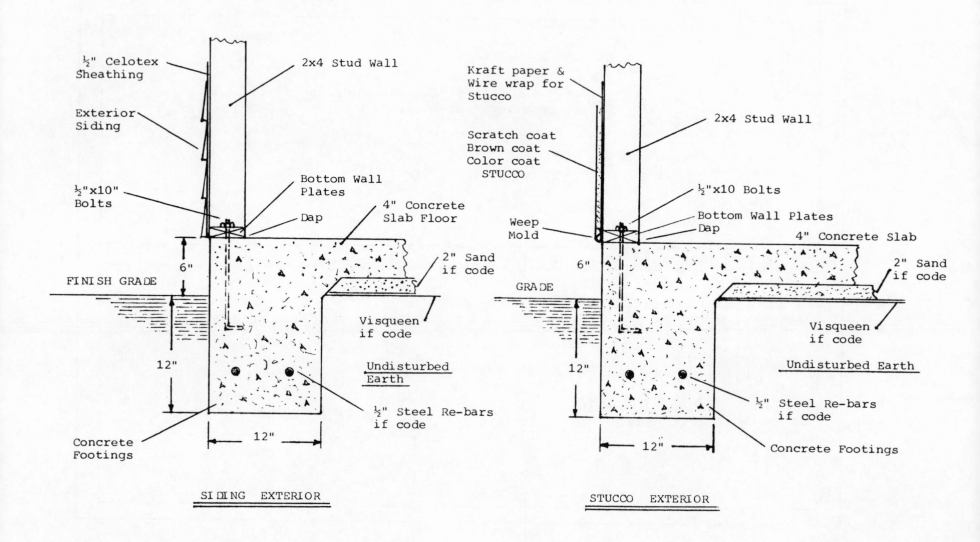

SIDING EXTERIOR

- ½" Celotex Sheathing
- Exterior Siding
- ½"x10" Bolts
- 2x4 Stud Wall
- Bottom Wall Plates
- Dap
- 4" Concrete Slab Floor
- 2" Sand if code
- FINISH GRADE
- 6"
- 12"
- Visqueen if code
- Undisturbed Earth
- Concrete Footings
- ½" Steel Re-bars if code
- 12"

STUCCO EXTERIOR

- Kraft paper & Wire wrap for Stucco
- Scratch coat Brown coat Color coat STUCCO
- Weep Mold
- 2x4 Stud Wall
- ½"x10 Bolts
- Bottom Wall Plates
- Dap
- 4" Concrete Slab
- 2" Sand if code
- GRADE
- 6"
- 12"
- Visqueen if code
- Undisturbed Earth
- ½" Steel Re-bars if code
- Concrete Footings
- 12"

CONCRETE SLAB FLOOR, COLD CLIMATE

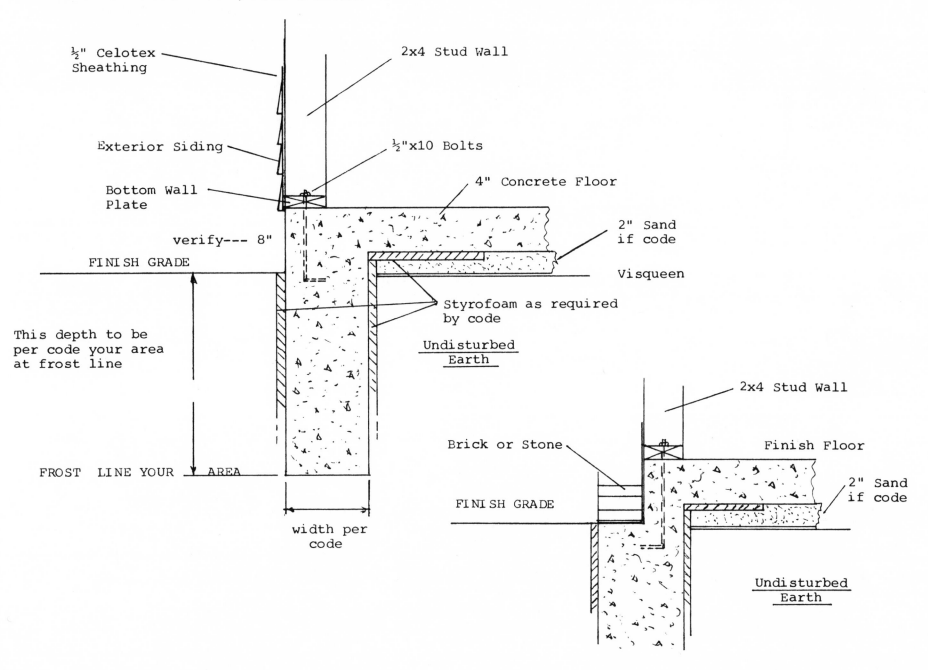

½" Celotex Sheathing

Exterior Siding

Bottom Wall Plate

2x4 Stud Wall

½"x10 Bolts

4" Concrete Floor

2" Sand if code

verify--- 8"

FINISH GRADE

Visqueen

Styrofoam as required by code

Undisturbed Earth

This depth to be per code your area at frost line

FROST LINE YOUR AREA

width per code

2x4 Stud Wall

Brick or Stone

Finish Floor

FINISH GRADE

2" Sand if code

Undisturbed Earth

GARAGE FOOTING DETAILS

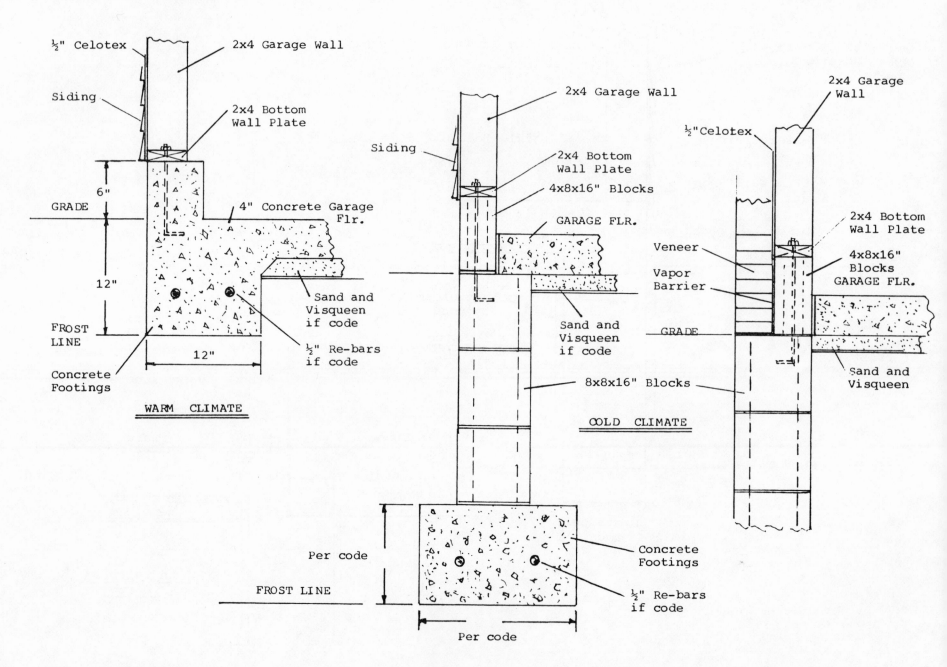

½" Celotex

Siding

2x4 Garage Wall

2x4 Bottom Wall Plate

GRADE

6"

4" Concrete Garage Flr.

12"

Sand and Visqueen if code

FROST LINE

12"

½" Re-bars if code

Concrete Footings

WARM CLIMATE

2x4 Garage Wall

Siding

2x4 Bottom Wall Plate

4x8x16" Blocks

GARAGE FLR.

Sand and Visqueen if code

8x8x16" Blocks

COLD CLIMATE

Per code

FROST LINE

Per code

Concrete Footings

½" Re-bars if code

2x4 Garage Wall

½"Celotex

2x4 Bottom Wall Plate

4x8x16" Blocks GARAGE FLR.

Veneer

Vapor Barrier

GRADE

Sand and Visqueen

WOOD FLOOR HOME WITH CRAWL SPACE

MATERIAL LIST

The following list of materials covers all items required from bottom of footings to subfloor level. Refer to notes below.

Description		Quantities	Cost
Concrete footings	cu. yd.	_____	_____
*Reinforcing-steel bars, ½″	lin. ft.	_____	_____
Concrete blocks (size)		_____	_____
Mortar and sand		_____	_____
*Vertical Reinforcing Bars, ⅜″	lin. ft.	_____	_____
*Concrete cap blocks (size)		_____	_____
Wall anchor bolts, ½″×10″		_____	_____
*Concrete for grouting		_____	_____
Concrete piers under beam	cu. yd.	_____	_____
8″×8″×16″ Concrete blocks for piers		_____	_____
*Visqueen vapor barrier	sq. ft.	_____	_____
Termite shields, 26 gauge	lin. ft.	_____	_____
2× Bottom plate	lin. ft.	_____	_____
2× Box end plate	lin. ft.	_____	_____

Labor Cost

Digging, forming, and pouring footings	_____	_____
Excavating crawl space	_____	_____
Laying blocks	_____	_____
*Reinforcing bars and grouting	_____	_____
*Visqueen and 2″ pea gravel	_____	_____
Forming and pouring garage floor	_____	_____
Floor-joist labor	_____	_____
Subfloor plumbing	_____	_____
*Forming and pouring driveway	_____	_____

*These items may not be required in your area.

Description		Quantities	Cost
2× Floor joists	lin. ft.	_____	_____
1″×4″ Joist "X" bracing	lin. ft.	_____	_____
6″×8″ Girder or I beam	lin. ft.	_____	_____
Termite shields, 26 gauge	lin. ft.	_____	_____
Subfloor, ½″ plywood	sq. ft.	_____	_____
50 lb. Box, 16p nails		_____	_____
50 lb. Box, 8p nails		_____	_____
*Pea-gravel crawl space	cu. yd.	_____	_____
Garage concrete footings	cu. yd.	_____	_____
Garage floor slab	cu. yd.	_____	_____
*Sand and Visqueen under slab		_____	_____
Wall anchor bolts, ½″×10″		_____	_____
*Wire mesh in slab	sq. ft.	_____	_____
*Reinforcing-steel bars, ½″	lin. ft.	_____	_____
*Concrete driveway	cu. yd.	_____	_____
16″×8″ Screened vents		_____	_____
Access window and well		_____	_____

Refer to Plumbing Chapter for details covering plumbing materials to be installed before the subfloor goes down.

Refer to Footing Section drawing covering proper details for your home. This detail is determined by type of exterior of your specific home: brick veneer, lap siding, stucco.

MATERIAL LIST: COST BREAKDOWN

The following paragraphs will cover all materials from bottom of footings to subfloor level. Each paragraph will guide you in figuring the quantities required to complete this portion of your new home. With these quantities, you can then obtain a total cost of materials needed to construct the foundation.

Excavating Crawl Space

Do not attempt this project by hand. Timewise it will be cheaper to hire someone to remove this earth. The excavator will have a transit with him. Set it up on the centerline of the street. Say that the height of instrument reading is 5′ and that

*These items may not be required in your area.

LOCATING DEPTH OF EXCAVATION FOR CRAWL SPACE OR BASEMENT

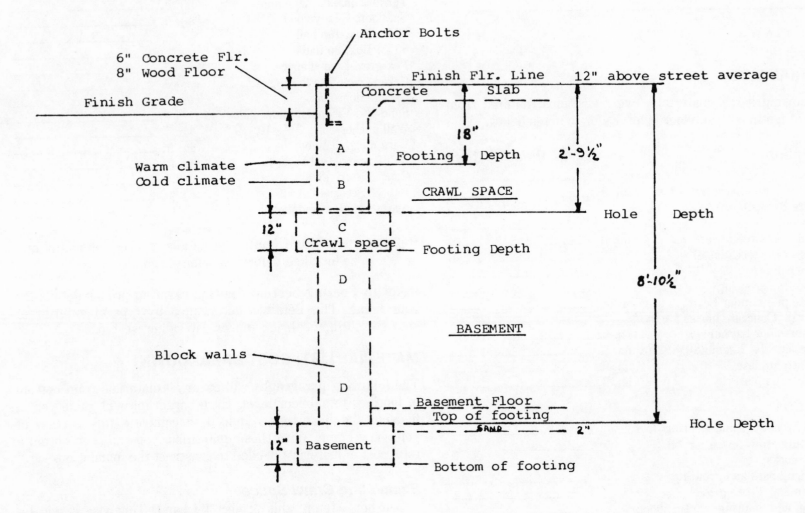

NOTE: *If elevation of existing land is below street level, earth removed from crawl space or basement area could fill in around home to proper height for finish grade. Less earth would be removed from the excavation area.*

Verify all dimensions and conditions on job site.

your code requires the floor to be 12″ above the street. The top of your finish floor stake should be driven down until you get a rod reading of 4′ with the rod on top of the stake. This will tell you how much earth must be removed down to the top of the concrete footings. Spread this earth out equally around the entire house for future ease of finish grading. The top of the concrete footings down from the top of the floor stake can be figured by adding up the height required for the subfloor, floor joist, and bottom plate, and the number of rows of blocks required to obtain proper crawl-space height per your code. Don't overlook the frostline depth in your area. If you need a 24″ crawl space and a 36″-deep frostline, you will be on the money with a 12″-deep footing and three rows of blocks on top. Verify all of this per code. Setting your floor line and footings will be one of your toughest projects; get it right when you start. You must also check at this time to see that your finish grade will come up to code. Review the sections on footings.

Concrete Footings

You are ready to hand-dig the trench for your footings. Check the width and depth shown on your drawings. After the footing trench is dug to the proper width, and very close to the proper depth, drive a stake into the center of the trench at one corner. The top of this stake will be driven down until the top is even with the top of the desired concrete footings. Check this measurement with the top of your finish-floor stake. This dimension was found when you started digging the crawl space. Review the above paragraph. You now have this one stake in the corner set exactly at the proper distance below the finish stake. Next, drive a stake about every 6′ in the center of the trench, completely around the trench. Using a good straight 8-foot 2 × 4, set the 2 × 4 on edge on top of first stake and on top of the second stake driven in. Set a 48″ level on the top edge of the 2 × 4, then drive the second stake down until the level reads perfectly even. Continue around the trench, from the second stake to the third, leveling each one as you go, until you are back to the starting stake. After all the stakes are set, with the tops all level, measure down from the top of each stake to see that you have room for the proper amount of con-

crete-footing thickness, as called for on your plans. At some stakes you may need to dig out some earth, while others may require some filling in to hold your desired footing depth. After this, you are ready to pour the concrete. Do not overlook your piers (as shown on page 41) down through the center of your home. Refer back to your blueprints for the size of these; perhaps 24″ × 24″ × 12″ deep. Remember, the top of these piers must be level with the top of your footings. Check back on your pier details and built-up beam details; note top of beam level with top of blocks.

Reinforcing-Steel Bars

In some areas, re-bars are not required in the footings. Check your plans to see if they have been called for in your home. If they are required, figure that you will need two ½″ bars completely around the footings. We are still using a 20′ × 40′ home as an example. This would be 2 times 120′ around the house, so you will need 240′ of ½″ re-bar.

Concrete Footings

Footings normally project about 6″ outside the block walls. (Refer to wall section drawings.) For your 20′ × 40′ home, you will probably use an 18″-wide by 12″-deep footing. Each running foot of this size footing is 120′ × 1⅓ cu. ft. of concrete. The distance around your home is 120′ × 1⅓ = 160 cu. ft. of concrete. (160 cu. ft. divided by 27 cu. ft. per cubic yard equals 6 cubic yards of concrete required to pour footings.) With footings poured, refer back to your batterboards. Set your lines on the batterboards to the outside of your footings, and then step in the proper distance from the outside edge of the footings to the outside edge of the block walls. Check your plans for this dimension. If this dimension is 6″, then set a nail in 6″ from the first nail on the top edge of the batterboards. Set these nails on all batterboards around all the sides of your home. Next, pull new lines to these new nails. Check these overall for exact dimensions. If you are building a home with siding exterior, these dimensions to the outside of the block walls will be the same as to the outside of your frame walls. Check this out very carefully before starting to lay blocks.

Concrete Wall Blocks

Check your plans for the size blocks called for under your home. Normally, these will be 8″×8″×16″, unless you will have a brick- or stone-veneer home. Refer to wall-section details and your plans. Since it is 120′ around your home, and the blocks are 16″ long, it will take about 90 blocks to complete one row around the house. If three rows are needed for a 24″ crawl space, you will require 270 blocks, plus center piers. When shopping for block prices, ask your local block supplier for the cost of mortar and sand to lay these blocks. He can deliver all of this at the same time. When you are ready to start laying blocks, call in someone experienced to help you set your corners. Check with your block supplier; he can find help for you. Get started off accurately and you can finish the job.

Masonry Wall Bolts

These bolts are ½″×10″ long with a square bend on one end and a nut and washer on the other. These bolts are set in concrete in the cores of the top row of blocks. Allow 2½″ of the threaded end to project up above the top of the block wall. These bolts secure the bottom plates, floor, and walls to the block walls and footings. Figure one bolt every 6′ around the entire top of the wall and two bolts at each corner—one 12″ each way from the corner—all set in concrete. (See detail drawings.) These bolts may not set directly on the center of block width; study this detail drawing closely.

Termite Shields

This item may be required in your area; check your plans. If they are required, your local sheet-metal shop can fabricate them for you in accordance with the code. Material used is 26-gauge galvanized sheet steel. When installing these, it will be necessary for you to mark and cut holes, allowing the shields to fit down over the masonry bolts. Lap the joints and tar each joint.

Bottom Plates

The bottom plates will usually be 14′-long 2×6's installed flat on top of the wall blocks and down over the masonry bolts. To mark and drill these plates, lay the 2×6 flat on the wall blocks on the inside of the bolts. With a square, mark a line on each side of each bolt; next, measure in from the outside of the block to the center of the bolt. Mark this distance in on the 2×6 and you will have the hole location. Fit 2×6 plates completely around the block walls. Mark and cut them all to fit before drilling a ¾″ hole at each bolt location.

Box End Plates

You now have the 2×6 bottom plates cut, drilled, and set down over the anchor bolts in the top of the block walls. Since most floor joists run from front to back, you will need box end plates only along the full length of the front and back of your home. Your model home is 40′ long. As a result, you need 80 linear feet of box end plates. These plates are always the same size as the floor joists. (Refer to the drawing showing box end plate installation.) You will note that the box end plate stands on edge on top of the outside edge of the 2×6 bottom plate. Nail the bottom plate to the bottom edge of the box end plate. This will form an L-shaped assembly ready to set down over the anchor bolts. Your floor joists will rest on top of the bottom plate and butt against the box end plate.

Floor Joists

To figure the number of floor joists required in your 40′-long home, start at one end of the home, laying the first floor joist flush with the outside block wall. Next, measure 16″ from the outside edge of the first floor joist to find the center of the second joist, and then 16″ center to center for each joist along the full 40′ span. You will need 31 floor joists across the front of the home and 31 joists across the back: 40′ divided by 16″. All walls running in the same direction as the floor joists must have two joists under each wall. Add these to the 62 joists already mentioned and you have just figured the number of joists needed. The length of the floor joist will be determined by the distance from front to back of your home. For a 20′ deep home, this would be 10′.

1×4 "X" Bracing

Refer to detail drawing on page 49 for complete details of this bracing. If 1×4 lumber is used, you should figure 48″ of

1×4 for each 16″ spacing. With 31 floor joists, you will have 30 spaces requiring "X" bracing, front and back: $2 \times 30 = 60$ spaces, times 4′ long, equals 240′ of 1×4 needed. You can purchase steel "X" braces which eliminate much time and labor. Check out your local suppliers.

Subfloor Plywood

In your model $20' \times 40'$ home, you have 800 square feet of subfloor area to cover. Plywood subfloor should always lie at right angles to the floor joists. Example: Your floor joists run from the front to the back of the home, so your $4' \times 8'$ sheets of plywood subfloor must run for 40′ along the width of the home. Use $\frac{1}{2}'' \times 4' \times 8'$ sheathing-grade plywood for the subfloor. Since you have a 40′ span to cover, 5 sheets 8′ long will do it. Since you have 20′ front to back to cover, you will need five 4′ sheets in that direction. Result: $5 \times 5 = 25$ sheets required. Buy two extra sheets; you will need them.

Note: Do not install subfloor until plumbing drains and lines are all roughed in. An inspection may be required in your area at this time. And you won't have to crawl around under the floor to install plumbing. Decide now what type of heating you will install. Will you need ducts under the floor?

Ground-Plumbing Drainage-System Layout

First, check with your building department to see if they must inspect the ground plumbing before the subfloor is installed, since it is quite a task to locate on the floor joists the exact spot where each pipe must come up through the floor. If they insist, take your plans and start by locating your walls on the edge of the box end plates. Allow 3½″ for wall thickness, plus your room dimensions.

Remember: All the drain lines coming up through the floor must come up inside a wall, except the toilet. Select your toilet at this time and ask the supplier the dimension from the wall to the center of the toilet flange. This will locate your toilet drain hole in the floor. Also, select your tub and/or shower at this time. Ask for drain location drawings. Just a tip—try to purchase all your plumbing materials and fixtures at a local plumb-ing wholesale house. You will save a lot of money. They can only say no—try it! Take your plans to them in the middle of the week, after lunch, when they are not too busy to help you. They will normally lay out your complete plumbing system, fitting by fitting, for you. If you cannot manage this, go to a plumber friend for help.

Some areas will allow a wood-floor home with crawl space to be framed, with all the walls standing in place and the sheathing on the roof, before the plumbing is roughed in. Let us say that you have the subfloor down and are ready to rough in plumbing.

Turn to the FRAMING CHAPTER (page 66) and study the methods of room layout, locating all the walls on the subfloor. With these lines snapped on the subfloor, you can then locate every fixture and cut holes through the floor, inside the walls. You can then run your drain lines to the holes in the floor.

Remember, all fixtures must have an S trap. These will all be under the floor, except those for the sinks. They will be in the cabinets. All fixtures must also have a vent pipe through the roof. All bathroom fixtures can be tied together with only one 3″ vent pipe projecting through the roof. Kitchen sink and washer will each have a 2″ vent pipe projecting through the roof. In most areas, your plumbing drain lines can be ABS black plastic pipe with glue joints. BE SURE TO PURCHASE ABS-TYPE GLUE. THIS IS A MUST.

On the Drain Layout Floor Plan, you will note that 3″ ABS is required for this hookup. At the house, you will also have a cleanout fitting with a screw cap. You should also have a cleanout under the bath lavatory and outside at the kitchen sink. All ABS drain lines should be installed with a slight slope toward the septic tank, perhaps ½″ per 10′. The top ends of the vent pipes should project through the roof about 16″. Each vent pipe should have a metal roof jack which slips down over the pipe and nails to the roof plywood. Apply a good amount of roof mastic or tar around each pipe and jack.

The above instructions should also be followed for basement-type homes. (See proper instructions for concrete slab floor homes.)

WOOD FLOOR CRAWL SPACE HOMES

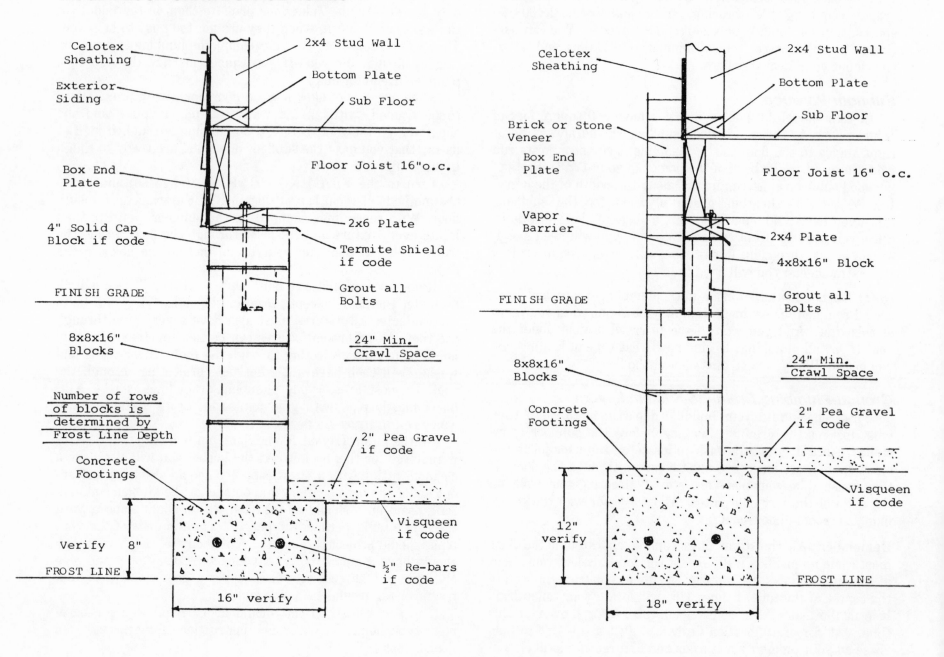

Celotex Sheathing

2x4 Stud Wall

Exterior Siding

Bottom Plate

Sub Floor

Box End Plate

Floor Joist 16"o.c.

4" Solid Cap Block if code

2x6 Plate

Termite Shield if code

FINISH GRADE

Grout all Bolts

8x8x16" Blocks

24" Min. Crawl Space

Number of rows of blocks is determined by Frost Line Depth

Concrete Footings

2" Pea Gravel if code

Verify 8"

Visqueen if code

FROST LINE

½" Re-bars if code

16" verify

Celotex Sheathing

2x4 Stud Wall

Bottom Plate

Brick or Stone Veneer

Box End Plate

Sub Floor

Floor Joist 16" o.c.

Vapor Barrier

2x4 Plate

4x8x16" Block

Grout all Bolts

FINISH GRADE

8x8x16" Blocks

24" Min. Crawl Space

Concrete Footings

2" Pea Gravel if code

12" verify

Visqueen if code

18" verify

FROST LINE

MASTER FOUNDATION PLAN FOR CRAWL SPACE HOME

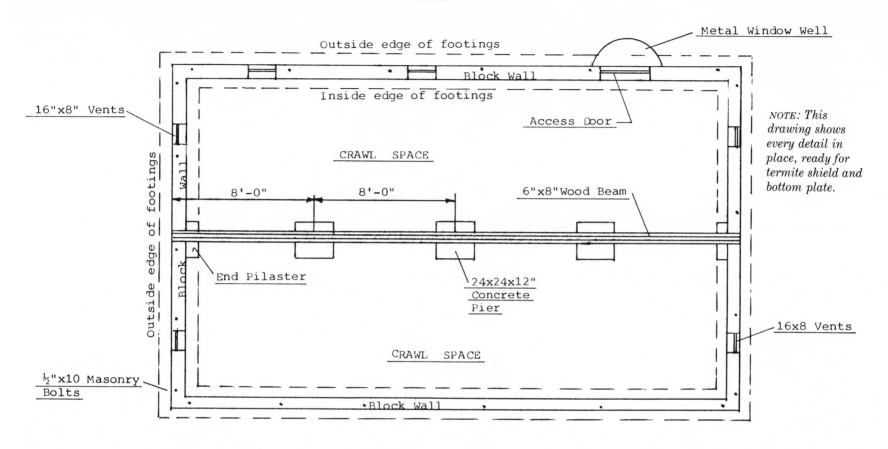

Metal Window Well

Outside edge of footings

Block Wall

Inside edge of footings

16"x8" Vents

Access Door

NOTE: *This drawing shows every detail in place, ready for termite shield and bottom plate.*

Outside edge of footings

Wall

Block

CRAWL SPACE

8'-0"   8'-0"

6"x8"Wood Beam

End Pilaster

24x24x12" Concrete Pier

16x8 Vents

½"x10 Masonry Bolts

CRAWL SPACE

Block Wall

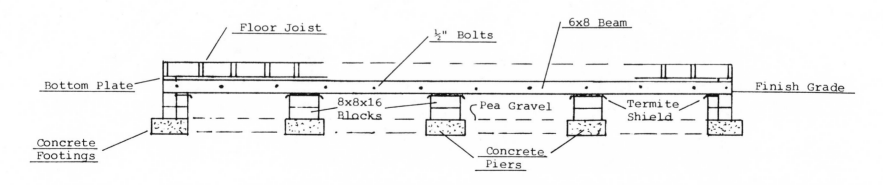

Floor Joist

½" Bolts

6x8 Beam

Bottom Plate

8x8x16 Blocks

Pea Gravel

Termite Shield

Finish Grade

Concrete Footings

Concrete Piers

FLOOR JOIST LAYOUT

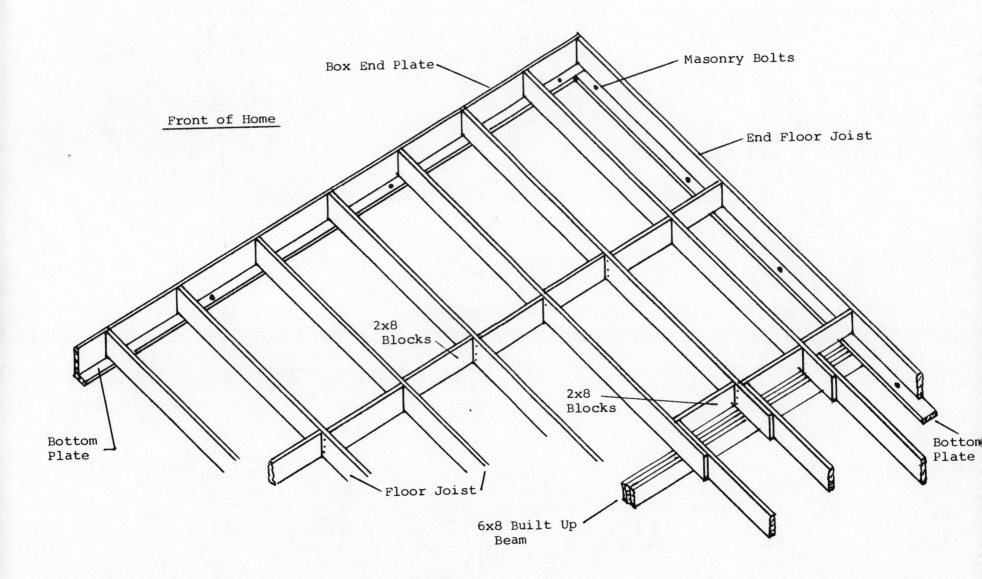

Box End Plate

Masonry Bolts

Front of Home

End Floor Joist

2x8
Blocks

2x8
Blocks

Bottom
Plate

Bottom
Plate

Floor Joist

6x8 Built Up
Beam

WOOD FLOOR HOME WITH BASEMENT

MATERIAL LIST

The following list of materials covers all items required from bottom of footings to subfloor level. The following items are listed as required during construction.

Description (House)		Quantities	Cost
Excavate basement	cu. yd.		
*Footing steel bars, ½″	lin. ft.		
Concrete footings	cu. yd.		
*Vertical reinforcing bars, ⅜″	lin. ft.		
Concrete blocks (size)			
Mortar and sand			
*Grouting concrete			
Basement windows and window wells			
Mastic mop and Visqueen	sq. ft.		
Wall anchor bolts, ½″ × 10″			
¾″ Rock for footing drains	cu. yd.		
4″ Tile footing drain	lin. ft.		
*Sump Pump Pit			
Visqueen under floor slab	sq. ft.		
4″ Concrete slab floor	cu. yd.		
Adjustable jack posts			
6×8 Girder or 6″ I beam	lin. ft.		
2× Bottom plates	lin. ft.		
2× Box end plates	lin. ft.		
2× Floor joist	lin. ft.		

*These items may not be required in your area.

		Quantities	Cost
1×4 "X" bracing	lin. ft.		
Basement stairs (see drawing)	lin. ft.		
Subfloor, ½″ plywood	sq. ft.		
Vapor barrier if brick veneer			

Description (Garage)		Quantities	Cost
Garage footings	cu. yd.		
*Visqueen vapor barrier	sq. ft.		
Concrete floor slab	cu. yd.		
Wall anchor bolts, ½″ × 10″			
*Wire mesh, reinforcing steel	sq. ft.		
*Reinforcing-steel bars, ½″	lin. ft.		
*2″ Sand under slab	cu. yd.		
*Concrete driveway	cu. yd.		
*Expansion joints	lin. ft.		

Labor Cost

Excavating basement		
Digging, forming, and pouring footings		
Block and grouting labor		
Mastic Mop and Visqueen labor		
Installing drain tile and rock		
Visqueen under basement and garage floor		
Forming and pouring basement floor		
Basement-stairs labor		
Floor-joist and subfloor labor		
Backfill outside basement walls		
Digging, forming, and pouring garage floor and footings		
Forming and pouring concrete driveway		

Refer to section drawings covering proper details for your home. **See Plumbing Chapter** for basement plumbing.

*These items may not be required in your area.

PLAN VIEW OF BASEMENT LAYOUT

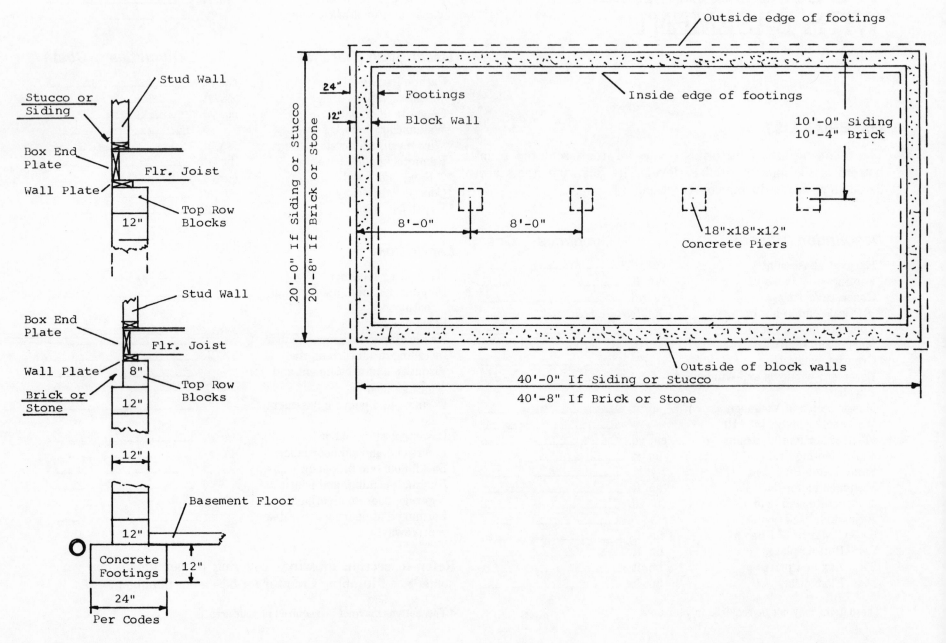

Stucco or Siding

Stud Wall

Box End Plate

Flr. Joist

Wall Plate

Top Row Blocks

12"

Stud Wall

Box End Plate

Flr. Joist

Wall Plate

8"

Brick or Stone

Top Row Blocks

12"

12"

Basement Floor

12"

12"

Concrete Footings

24"

Per Codes

Outside edge of footings

24"

12"

Footings

Inside edge of footings

Block Wall

10'-0" Siding
10'-4" Brick

20'-0" If Siding or Stucco
20'-8" If Brick or Stone

8'-0"

8'-0"

18"x18"x12"
Concrete Piers

Outside of block walls

40'-0" If Siding or Stucco
40'-8" If Brick or Stone

BASEMENT WALL SECTION

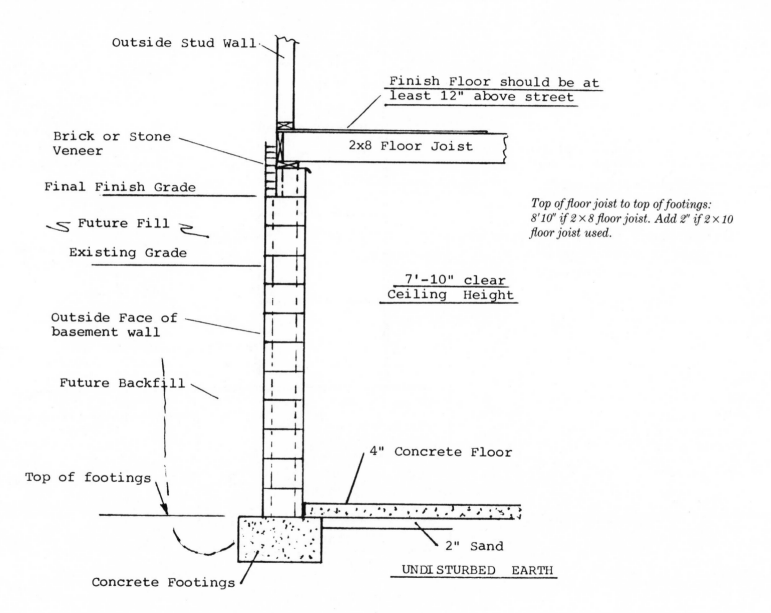

Outside Stud Wall

Finish Floor should be at least 12" above street

Brick or Stone Veneer

2x8 Floor Joist

Final Finish Grade

Top of floor joist to top of footings: 8'10" if 2 × 8 floor joist. Add 2" if 2 × 10 floor joist used.

Future Fill

Existing Grade

7'-10" clear Ceiling Height

Outside Face of basement wall

Future Backfill

4" Concrete Floor

Top of footings

2" Sand

Concrete Footings

UNDISTURBED EARTH

45

LOWER BASEMENT WALL DETAILS

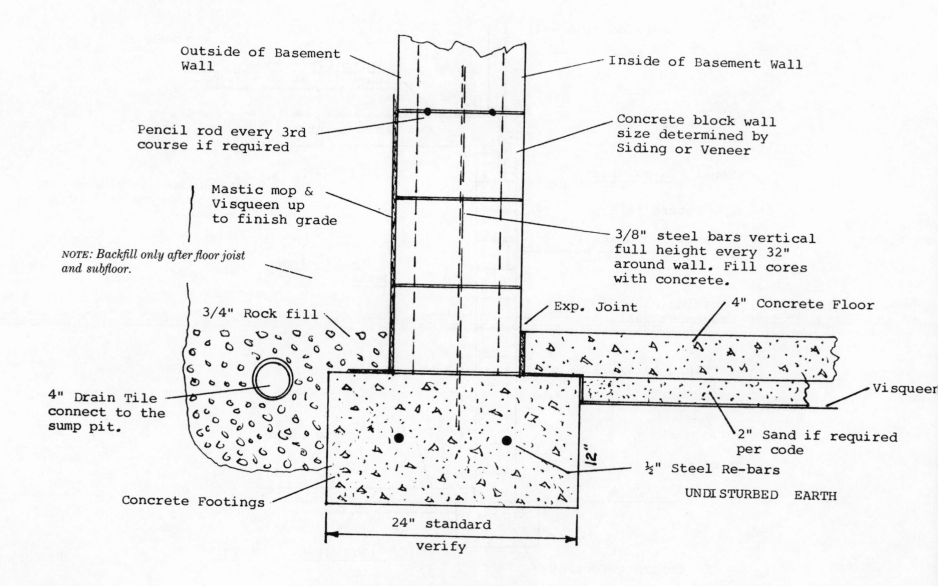

Outside of Basement Wall

Inside of Basement Wall

Pencil rod every 3rd course if required

Concrete block wall size determined by Siding or Veneer

Mastic mop & Visqueen up to finish grade

NOTE: Backfill only after floor joist and subfloor.

3/8" steel bars vertical full height every 32" around wall. Fill cores with concrete.

3/4" Rock fill

Exp. Joint

4" Concrete Floor

4" Drain Tile connect to the sump pit.

Visqueen

2" Sand if required per code

12"

½" Steel Re-bars

Concrete Footings

UNDISTURBED EARTH

24" standard
verify

TOP OF BASEMENT WALL DETAILS

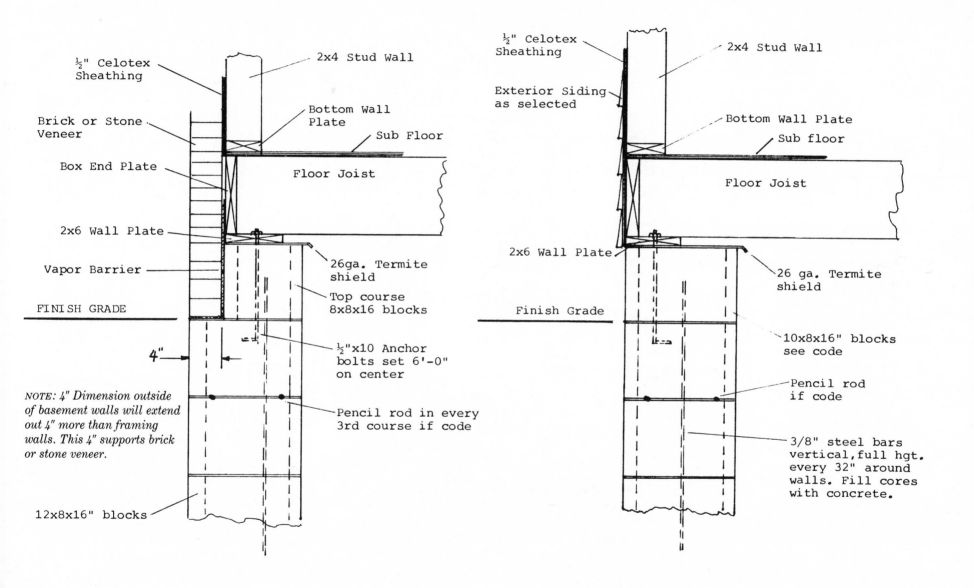

½" Celotex Sheathing

Brick or Stone Veneer

Box End Plate

2x6 Wall Plate

Vapor Barrier

FINISH GRADE

4"

NOTE: 4" Dimension outside of basement walls will extend out 4" more than framing walls. This 4" supports brick or stone veneer.

12x8x16" blocks

2x4 Stud Wall

Bottom Wall Plate

Sub Floor

Floor Joist

26ga. Termite shield

Top course 8x8x16 blocks

½"x10 Anchor bolts set 6'-0" on center

Pencil rod in every 3rd course if code

½" Celotex Sheathing

Exterior Siding as selected

2x6 Wall Plate

Finish Grade

2x4 Stud Wall

Bottom Wall Plate

Sub floor

Floor Joist

26 ga. Termite shield

10x8x16" blocks see code

Pencil rod if code

3/8" steel bars vertical, full hgt. every 32" around walls. Fill cores with concrete.

SECTION THROUGH COMPLETE BASEMENT WALL

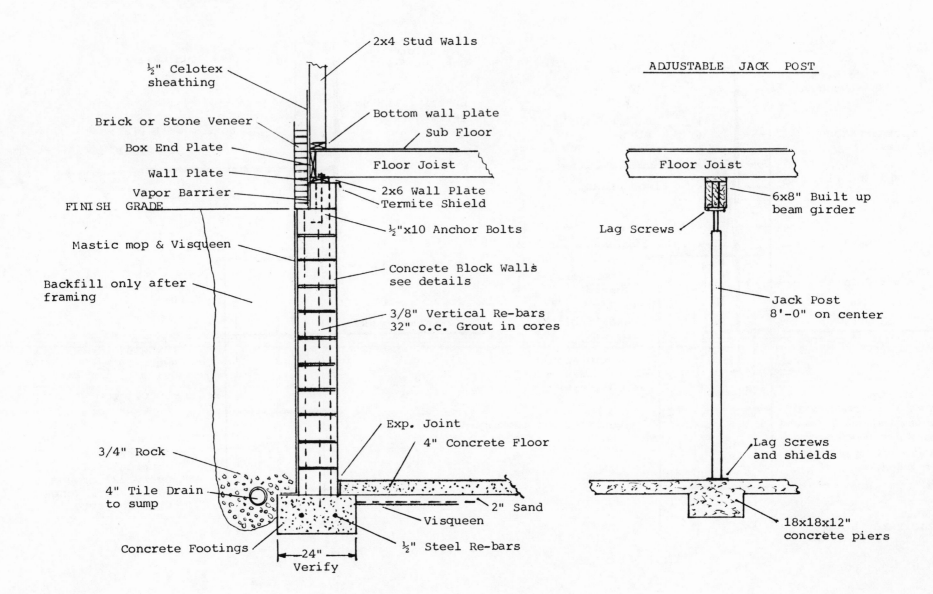

2x4 Stud Walls

½" Celotex sheathing

Brick or Stone Veneer

Box End Plate

Wall Plate

Vapor Barrier

FINISH GRADE

Mastic mop & Visqueen

Backfill only after framing

3/4" Rock

4" Tile Drain to sump

Concrete Footings

24" Verify

Bottom wall plate

Sub Floor

Floor Joist

2x6 Wall Plate
Termite Shield

½"x10 Anchor Bolts

Concrete Block Walls see details

3/8" Vertical Re-bars 32" o.c. Grout in cores

Exp. Joint

4" Concrete Floor

2" Sand

Visqueen

½" Steel Re-bars

ADJUSTABLE JACK POST

Floor Joist

6x8" Built up beam girder

Lag Screws

Jack Post 8'-0" on center

Lag Screws and shields

18x18x12" concrete piers

48

SECTION VIEW THROUGH BASEMENT

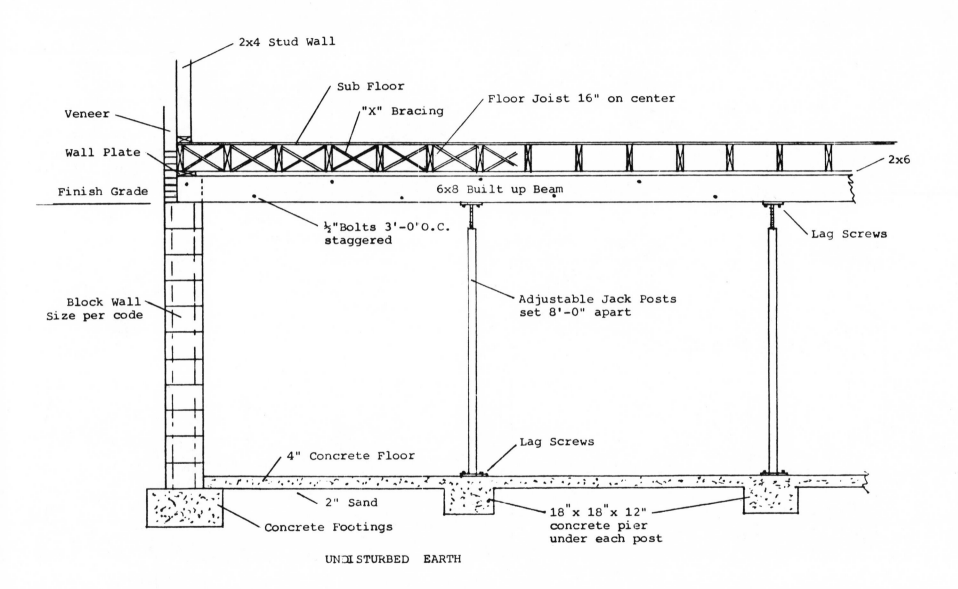

2x4 Stud Wall

Sub Floor

"X" Bracing

Floor Joist 16" on center

Veneer

Wall Plate

Finish Grade

2x6

6x8 Built up Beam

½"Bolts 3'-0'O.C. staggered

Lag Screws

Block Wall Size per code

Adjustable Jack Posts set 8'-0" apart

4" Concrete Floor

Lag Screws

2" Sand

Concrete Footings

18"x 18"x 12" concrete pier under each post

UNDISTURBED EARTH

BASEMENT OVERHEAD BEAM DETAILS

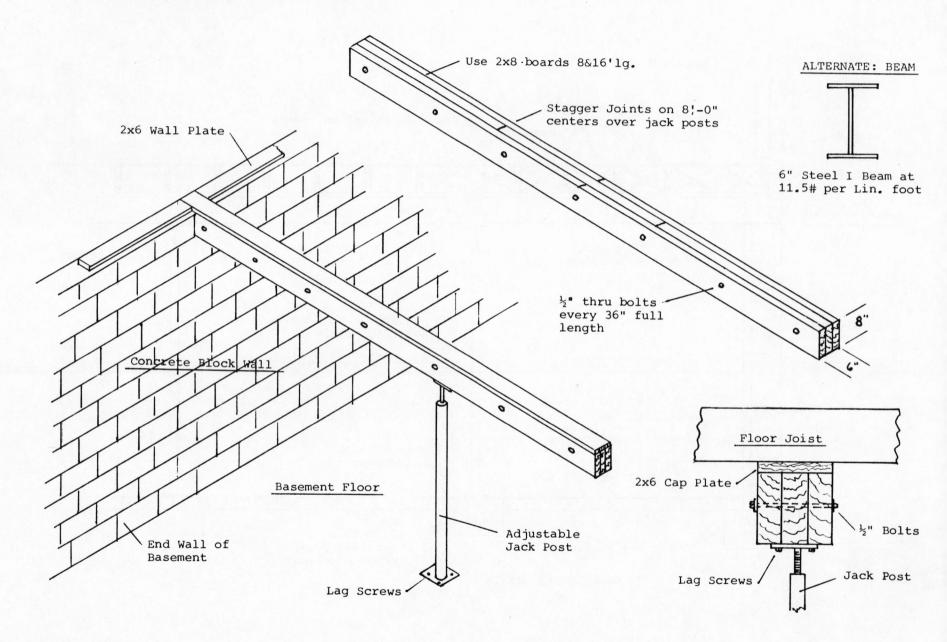

Use 2x8 boards 8&16'lg.

Stagger Joints on 8'-0" centers over jack posts

2x6 Wall Plate

ALTERNATE: BEAM

6" Steel I Beam at 11.5# per Lin. foot

½" thru bolts every 36" full length

8"

6"

Concrete Block Wall

Basement Floor

Floor Joist

2x6 Cap Plate

½" Bolts

End Wall of Basement

Adjustable Jack Post

Lag Screws

Lag Screws

Jack Post

SUGGESTED BASEMENT WINDOW LAYOUT

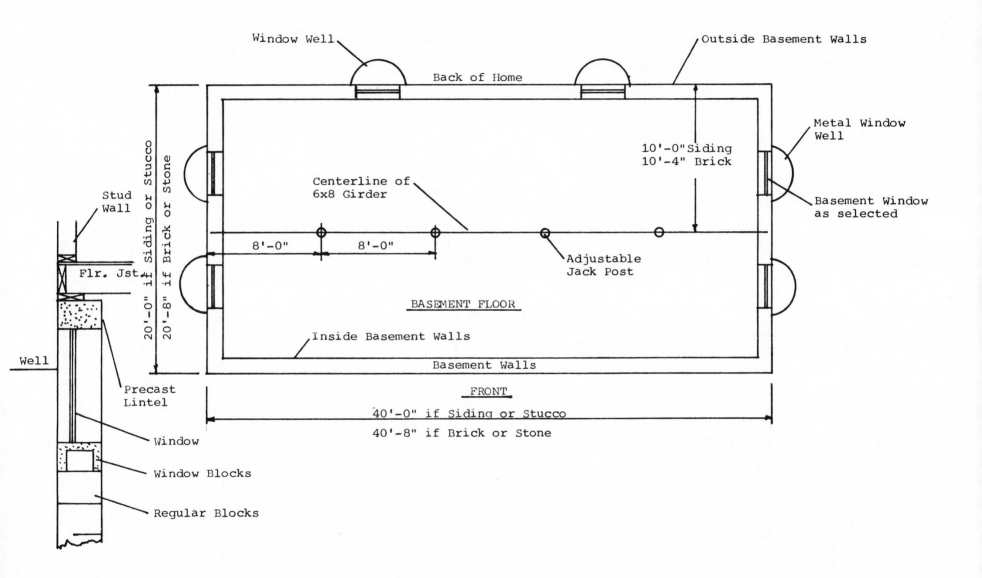

Window Well

Outside Basement Walls

Back of Home

Metal Window Well

10'-0" Siding
10'-4" Brick

Basement Window as selected

Centerline of 6x8 Girder

Stud Wall

Flr. Jst.

8'-0"   8'-0"

20'-0" if Siding or Stucco
20'-8" if Brick or Stone

Adjustable Jack Post

BASEMENT FLOOR

Well

Inside Basement Walls

Precast Lintel

Basement Walls

Window

FRONT

40'-0" if Siding or Stucco

40'-8" if Brick or Stone

Window Blocks

Regular Blocks

BASEMENT STAIRS DETAILS

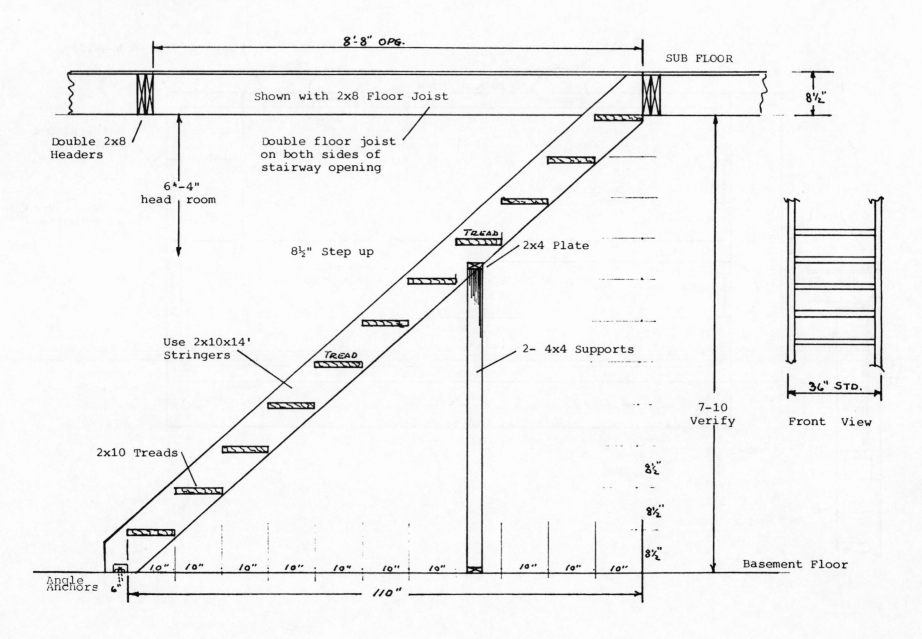

8'-8" OPG.

SUB FLOOR

8½"

Double 2x8 Headers

Shown with 2x8 Floor Joist

Double floor joist on both sides of stairway opening

6"-4" head room

8½" Step up

TREAD

2x4 Plate

Use 2x10x14' Stringers

TREAD

2- 4x4 Supports

2x10 Treads

7-10 Verify

36" STD.

Front View

8½"

8½"

8½"

Angle Anchors

6"

10" 10" 10" 10" 10" 10" 10"   10" 10" 10"

110"

Basement Floor

52

WATER-SUPPLY PLUMBING LAYOUT FOR CONCRETE FLOOR

LAYOUT INSTRUCTIONS

After the footings are dug for a concrete slab floor, the next step is to install the ground-plumbing drainage system. Refer to the proper drawings on page 58 covering this phase.

All fixtures must be located at a proper distance from the forms and must be exact. With the toilet flange set even with the top of the floor and the toilet bend in place, all dimensions start from this point or depth. Our 3″ ABS will run from toilet to cleanout and then to septic tank out at least 5′ from the house; I suggest 10′. From the toilet, go 3″ over to and up with the 3″ vent. Out of the end of this 3″ ell, continue with the 2″ ABS over to the lavatory, washer, and kitchen sink, ending with the 2″ cleanout at the back of your home.

Catch all vent risers as you come to them. Tub, toilet, and lavatory should be on the same 3″ vent. The washer and kitchen sink each have a 2″ vent. Remember, from the toilet to the kitchen sink you should have a constant slight rise to direct the flow to the septic tank.

At each bathtub location, glue in the S trap with the top no higher than the concrete floor. (All other S traps will be above the slab.) Build a 12″ square box of 2 × 4's and set it down over the tub drain and toilet flange. The top of the box should be flush with the top of the slab. This will help keep the concrete away from the drains.

Next, glue in a short ABS pipe riser at each fixture to extend about 24″ above the concrete floor. Do not cover this drainage system. It must first be inspected with all waterlines installed. Cap the ends and fill the entire system with water for a leak test.

Water-Supply Plumbing

With open ditches and piles of dirt everywhere, you are now ready to install your hot- and cold-water lines under the slab. All of these lines should be set 3″ or 4″ below the bottom of the slab.

First, drive a wooden stake into the ground within the 3½″-wall thickness at each fixture location: at the valve at the end of the tub, to the left of the toilet, in the wall at the water heater, in the wall at the center of the washer, and in the wall at the center of the kitchen sink. Recheck these dimensions closely. Don't give up. Normally you will have a 1″ waterline coming in from the street. This is usually a 1″ galvanized pipe buried down to the frostline in your area. At the house, install a ¾″ shutoff valve to stop all water from coming into the house.

Next, a ¾″ pressure regulator. At this point you can tee off to the hose bibb by the bathroom. It is not necessary to set the shutoff valve or pressure regulator at this time.

Leave 24″ of your ¾″ soft copper tubing sticking out in front of the house; use duct tape and seal the end to keep dirt out.

Next, unroll the ¾″ soft copper tubing back to the stake at the water-heater location. Make an easy, gentle bend 90 degrees up the side of the stake. Cut the tubing off 24″ above the floor and tape it to the stake using the duct tape. Tape the end shut.

Cold-Water Lines

You will now install the cold-water lines. Always tape the cold lines to the right side of each stake as you face the fixture: hot on the left, cold on the right. Using a coil of ½″ soft copper tubing, which comes in 60′ coils, make another gentle 90-degree bend with a 24″-tail. Tape this tail to the ¾″ tail sticking up at the water heater. Uncoil the roll over to the washer, make a gentle 90-degree bend with a 24″-tail, cut the tubing, and tape it to the right side of the stake.

Uncoil another run from the washer to the right side of the stake to the kitchen sink, with a 24"-tail.

No joints are allowed in the waterlines under the concrete slab. Cap all open ends of all waterlines as you install them. A pipe filled with dirt will cause you many problems. If you keep all cold-water lines on the right side of the stakes and all hot lines on the left side of the stakes, there will be less chance of fouling up when the lines are tied together.

Next, run a ½" line from the cold water at the water heater to the right-side stake at the bathroom lavatory. Make another 24"-tail and gentle bend. If you should happen to flatten or close the tube while bending it, cut it off and make a new bend. *Never hide it*—you won't get any water through it if you do. Make all cuts of the tubing with a regular tubing cutter and sand the ends with emery cloth for 1" back. The connection from the lavatory to the toilet can be made later in the stud wall. Uncoil another ½" line from the toilet under the slab over to the tub; make the same tail and bend. All of these cold-water lines must lie in a shallow trench 3" or 4" deep. Leave the trench open for the inspector to check. The lines must lie 3" or 4" below the bottom of the concrete slab.

Hot-Water Lines

You are now ready to install the hot-water lines. Starting at the water heater, lay a ½" line under the slab over to the bathroom lavatory, coming up inside the wall. Make the same gentle bend and 24" tail. Tape this tail to the left side of the stake. Run another line from the lavatory to the left side of the stake at the tub; same gentle bend and tail.

Next, starting at the water heater, run a ½" line under the slab over to the left side of the stake at the washer. Same bend and tail. From the washer, run another line to the left of the kitchen sink.

You now have all your fixture stakes set, and all hot and cold lines laid in a shallow trench below the bottom of the concrete slab. All of your hot-water lines are anchored on the left sides of the stakes, and all cold-water lines anchored on the right sides of the stakes. You are now ready to solder or sweat

the joints, as shown on pages 61–64. *Study these sketches carefully before you start.*

Equipment and Tools Needed to Complete This Project

You already have your tubing cutter. You will now need a small bottle-gas hand-torch kit for heating the joints. Also, a small can of paste flux, a brush for the flux, a roll of acid-core wire solder, and a roll of plumber's 1½"-wide emery cloth. After making a cut from your copper tubing, take the emery cloth and polish the end of the tubing that will be soldered. Shine the tubing back about 1".

Next, coat this clean area with paste flux, using a small flux brush. If you are going to solder an elbow or any other fitting onto this tubing, use the flux to coat the inside of the fitting where the tubing fits in.

Next, slip the fitting firmly over the tubing. Be sure it is all the way in. You are now ready to sweat a joint. Fire up your torch—not too high. You will soon get acquainted with your torch. *Remember, it gets red hot* and it will burn anything. Lay your fitting and tubing on a flat surface and slowly move your flame back and forth over the area to be soldered. Roll your wire solder out about 12" and touch the end of the solder to the heated joint. Keep the flame passing back and forth and keep touching the solder to the joint. When the copper is hot enough, the wire solder will melt instantly. When the solder melts, pull the flame away slightly and feed the solder into the joint, not too much solder but enough to see it completely around the joint.

Don't let different fittings and pipes touch when connecting several in a close area, as may happen with cold-water lines at the water heater. See the drawing on page 56. It is best to cut and fit all pipes and fittings before soldering. Flux and clean all joints and fittings as you assemble them. Then everything will fit and you will be ready to sweat them together.

NOTE: All pipes that will come up through the slab can be moved slightly as long as they stay within the wall thickness. Move them slightly for ease of assembly. Example: At the water-heater cold-water assembly, you will have eight joints to sweat within a 6″ or 8″ distance. If you keep too much heat in the wrong place too long, you will melt the joint you have just finished. Learn to heat your pipes at a slight distance from the fitting.

You are about to sweat! It will take some practice before you become an expert. Sweat all the cold-water lines first. Stay with the cold-water lines until they are complete. At the front of the house where the water comes in, you left 24″ of ¾″ copper sticking outside the forms. Clean up the end of this pipe with the paste flux and emery cloth and sweat on a ¾″ male-thread copper adapter for hookup to a lawn hose. The hose will be used for a water-pressure test of all the copper lines to catch any leaks. You cannot sweat a joint with water in the pipe.

On the water-connection sketches of each fixture (pages 61–64), you will note a STOP line on every fixture sketch. Stop your pipes at this point and sweat on a pipe cap for the water test. After the slab is poured and the framing is up, cut this cap off with a cutter. Then sweat on a coupling and go on up with the pipes as shown in the drawing.

Next, at your water-heater location, sweat a temporary connection from the cold system over to the hot system. This is needed to test the hot lines for leaks along with the cold ones.

You are now ready to call for a footing and ground-plumbing inspection. It looks like the devil, but the inspector must see everything.

Great! Two or three little things to do, but you passed your first inspection. If you don't understand something the inspector says to you, ask him to explain it. Your next call for inspection will be after the plywood is on the roof. However, some areas must be reinspected just before you pour the concrete slab. Check this out with the inspector.

You are now ready to fill in the ditches to cover all the waterlines and drain lines. Spread the soil out smoothly and hold it 3½″ below the top of the forms. This is to allow for the proper thickness of the concrete slab. Use a good straight 2 × 4 on edge. This is the thickness your concrete will be. Clean out your footing trenches to be ready for the concrete. Are you required to use ½″ steel bars in your footings? *Check your plans.* You should sprinkle the entire area with water over and over to compact the soil. Fill in a little if necessary.

This is it. You are ready to pour the concrete. You must enlist a lot of help to handle this amount of concrete at one time. Call in a concrete man. Have him check you out in every detail. He will have two or three men finish a beautiful slab for you. Don't forget the 12″-square boxes around the tub and toilet drains. Do you have the ½″ × 10″ anchor bolts?

MASTER PLUMBING LAYOUT FOR TYPICAL HOME

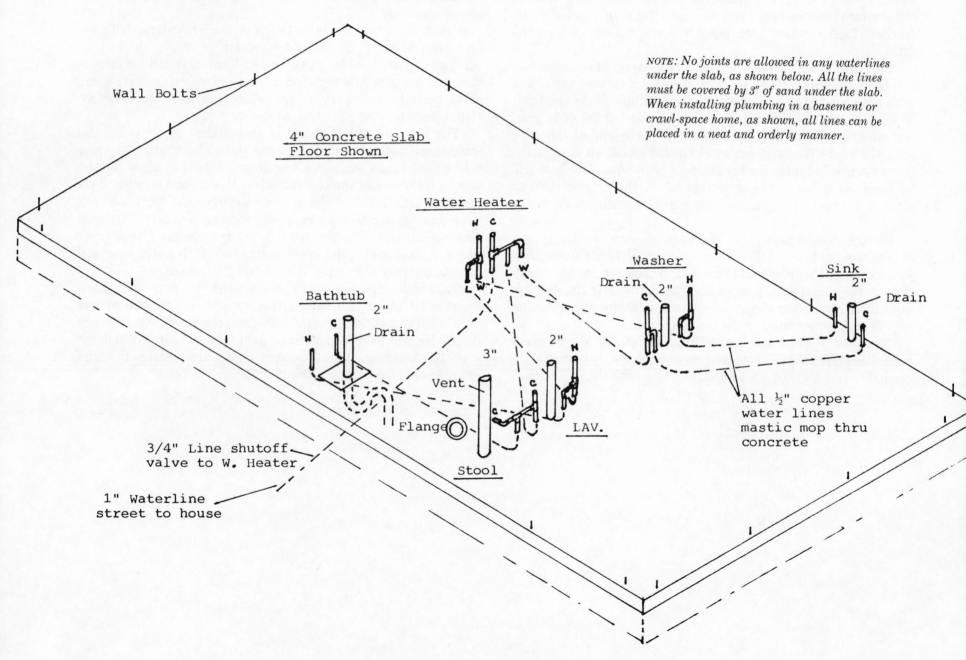

Wall Bolts

4" Concrete Slab
Floor Shown

NOTE: *No joints are allowed in any waterlines under the slab, as shown below. All the lines must be covered by 3" of sand under the slab. When installing plumbing in a basement or crawl-space home, as shown, all lines can be placed in a neat and orderly manner.*

Water Heater

Washer

Sink
2"

Bathtub
2"
Drain

Drain
2"

Drain

Vent

2"

3"

LAV.

Flange

All ½" copper
water lines
mastic mop thru
concrete

3/4" Line shutoff
valve to W. Heater

Stool

1" Waterline
street to house

56

TYPICAL PLUMBING DRAIN LAYOUT

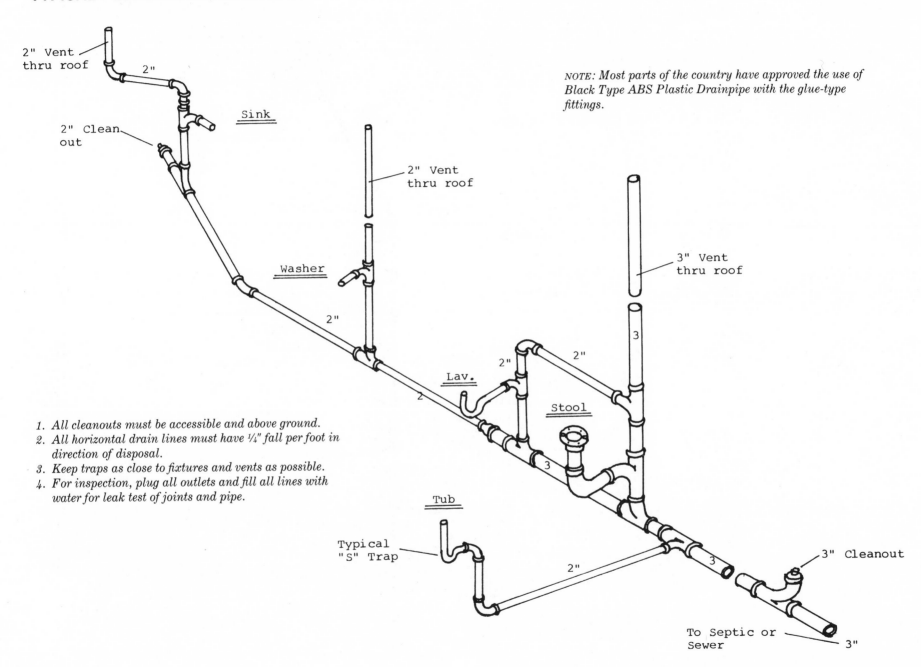

2" Vent thru roof

2"

Sink

2" Clean out

NOTE: *Most parts of the country have approved the use of Black Type ABS Plastic Drainpipe with the glue-type fittings.*

2" Vent thru roof

3" Vent thru roof

Washer

2"

3

2"

Lav.

2"

2"

2

3

Stool

1. *All cleanouts must be accessible and above ground.*
2. *All horizontal drain lines must have ¼" fall per foot in direction of disposal.*
3. *Keep traps as close to fixtures and vents as possible.*
4. *For inspection, plug all outlets and fill all lines with water for leak test of joints and pipe.*

Tub

Typical "S" Trap

3

3" Cleanout

2"

To Septic or Sewer

3"

GROUND-PLUMBING DRAINAGE SYSTEM LAYOUT

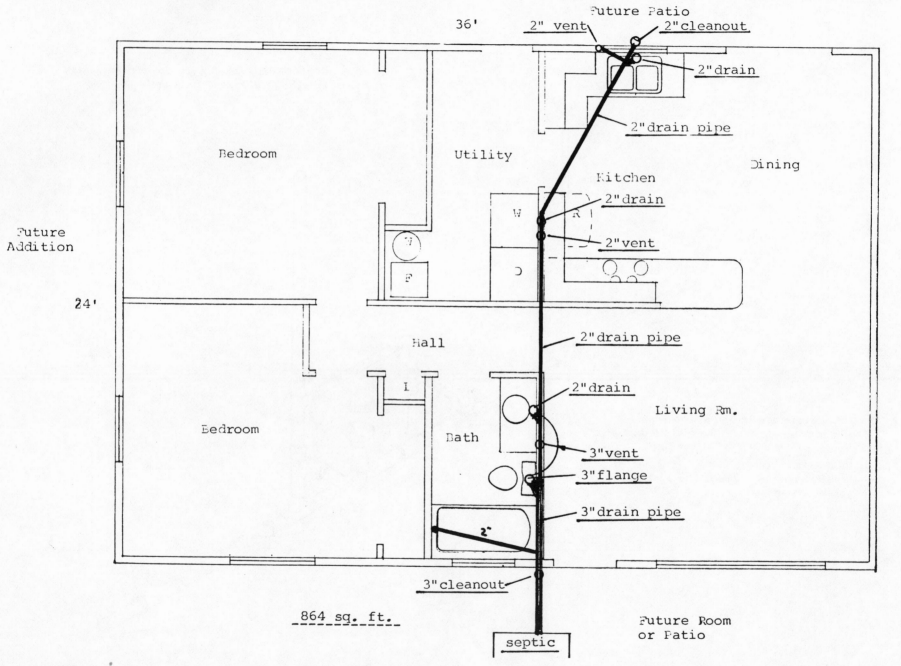

Future Patio

36'

2" vent   2"cleanout

2"drain

2"drain pipe

Bedroom   Utility   Dining

Kitchen

2"drain

W   R

2"vent

Future
Addition

Future
Garage

24'

Hall

2"drain pipe

I

2"drain

Living Rm.

Bedroom

3"vent

3"flange

Bath

3"drain pipe

2"

3"cleanout

864 sq. ft.

Future Room
or Patio

septic

WATER-SUPPLY PLUMBING LAYOUT FOR CONCRETE FLOOR

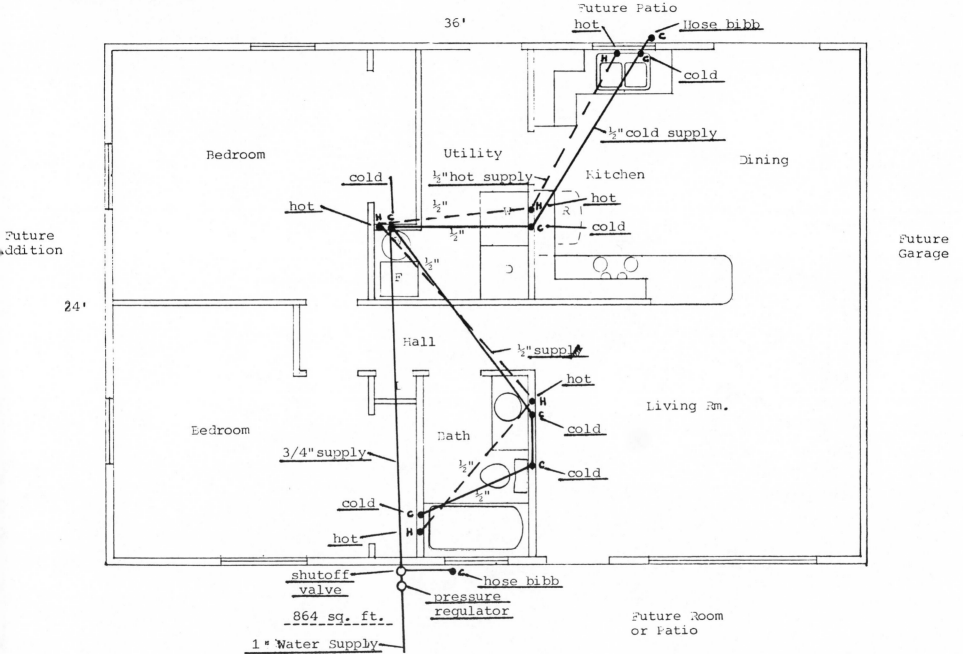

Future Patio

hot   Hose bibb

cold

½" cold supply

Bedroom   Utility   Kitchen   Dining

cold   ½" hot supply

hot   ½"   W   H   hot

H C   R   C   cold

½"   ½"

F

24'

Hall

½" supply

hot

H

Living Rm.

C   cold

Bedroom

3/4" supply   ½"   C   cold

Bath

cold   ½"

C

hot   H

shutoff   C   hose bibb
valve

pressure
regulator

864 sq. ft.

Future Room
or Patio

1" Water Supply

Future
Addition

Future
Garage

36'

WATER-SUPPLY PLUMBING LAYOUT FOR CRAWL-SPACE OR BASEMENT HOME

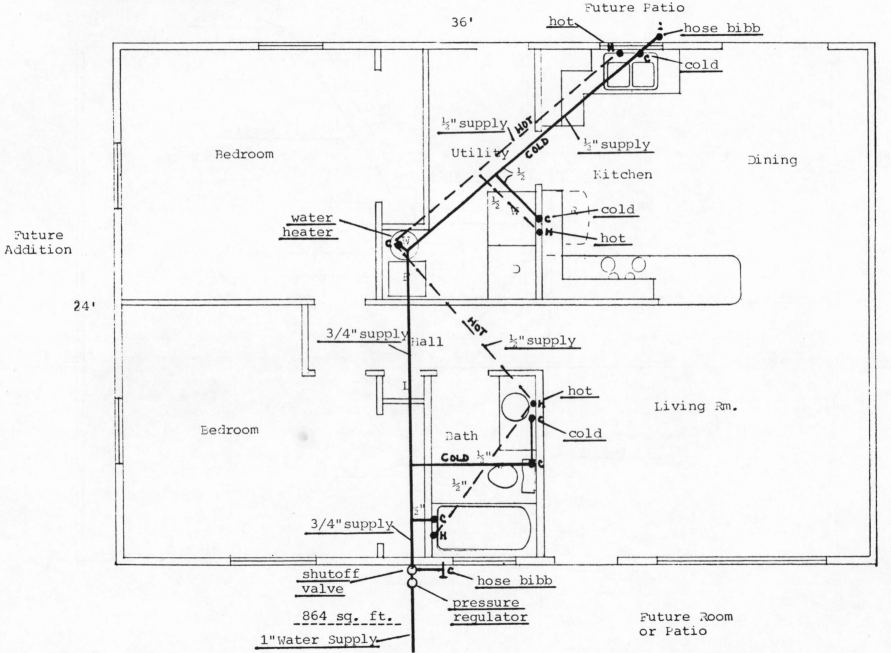

Future Patio

hot

hose bibb

cold

36'

½" supply

HOT

COLD

½" supply

Utility

Kitchen

Dining

½

½

¾

cold

hot

water
heater

Future
Addition

Futu
Gara

24'

3/4" supply

Hall

HOT

½" supply

hot

Living Rm.

cold

Bedroom

Bath

COLD ½"

cold

½"

3/4" supply

shutoff
valve

hose bibb

864 sq. ft.

pressure
regulator

Future Room
or Patio

1" Water Supply

COLD-WATER FIXTURE CONNECTIONS

NOTE: The connections shown below are for a concrete slab floor. However, the same connections will be required for crawl-space or basement homes, except that the water heater would be set in the basement along with the furnace, washer, and dryer. The utility room would become a stairway down to the basement. In crawl-space and basement homes, some areas will allow the use of a PVC schedule-40 plastic pipe for cold-water lines and a special PVC pipe for hot-water lines. Check this out; it could be quicker and cheaper than the alternatives. Install plumbing lines up to the stop lines (as shown on the sketches) before pouring the concrete slab. Place a sweat cap on the end of every line at the stop line. Complete plumbing above the stop line only after the walls are up.

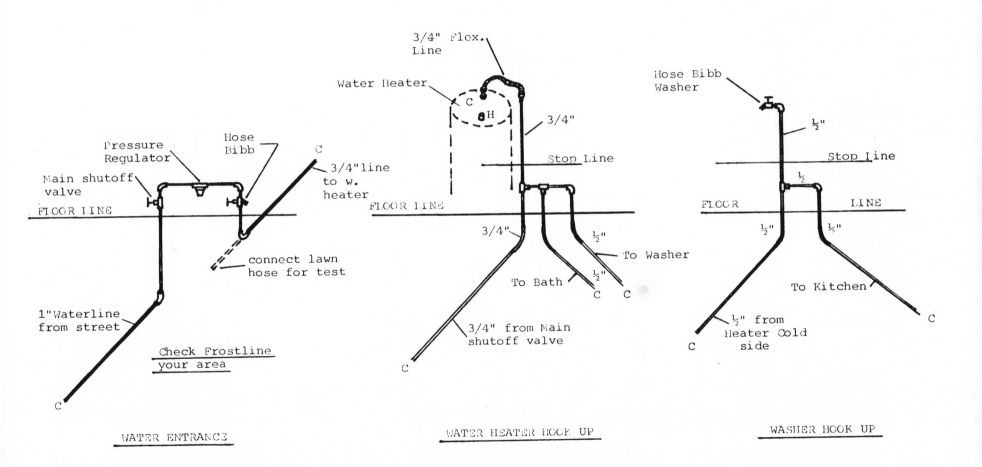

WATER ENTRANCE

WATER HEATER HOOK UP

WASHER HOOK UP

COLD-WATER FIXTURE CONNECTIONS (CONT.)

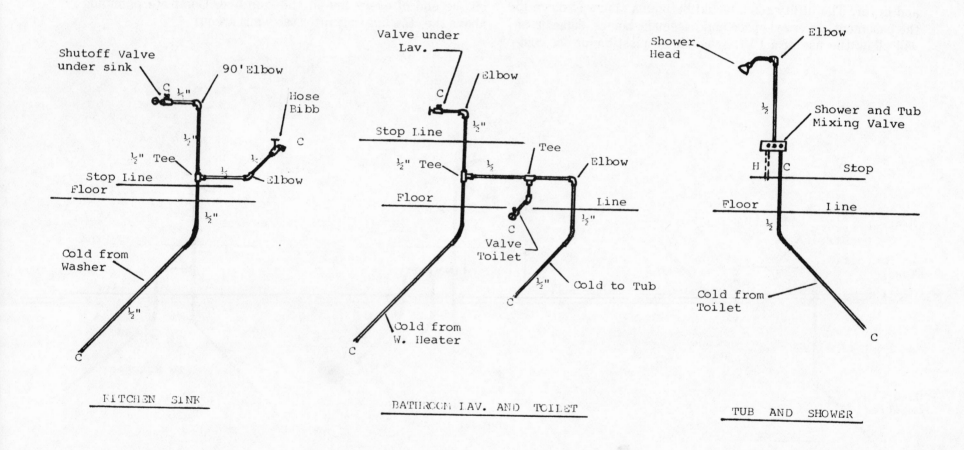

KITCHEN SINK

BATHROOM LAV. AND TOILET

TUB AND SHOWER

HOT-WATER FIXTURE CONNECTIONS

NOTE: The connections shown below are for a concrete slab floor. However, the same connections will be required for crawl-space or basement homes, except the water heater would be set in the basement along with the furnace, washer, and dryer. The utility room would become a stairway down to the basement. In crawl-space and basement homes, some areas will allow the use of a PVC schedule-40 plastic pipe for cold-water lines and a special PVC pipe for hot-water lines. Check this out; it could be quicker and cheaper than the alternatives. Install plumbing lines up to the stop lines (as shown on the sketches) before pouring the concrete slab. Place a sweat cap on the end of every line at the stop line. Complete plumbing above stop line only after walls are up.

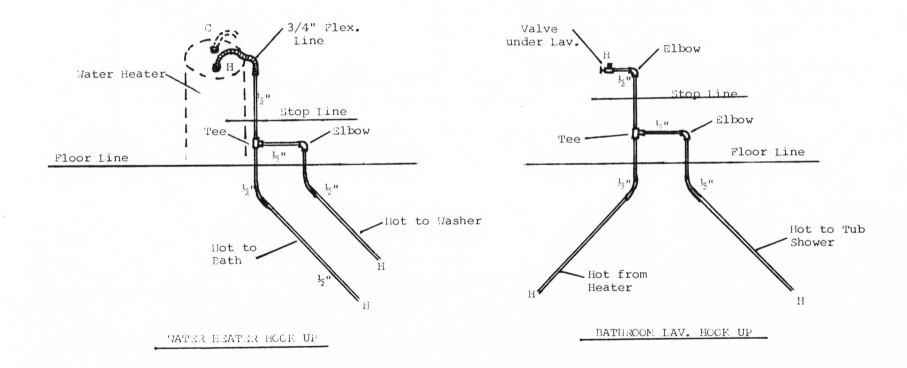

WATER HEATER HOOK UP

BATHROOM LAV. HOOK UP

HOT-WATER FIXTURE CONNECTIONS (CONT.)

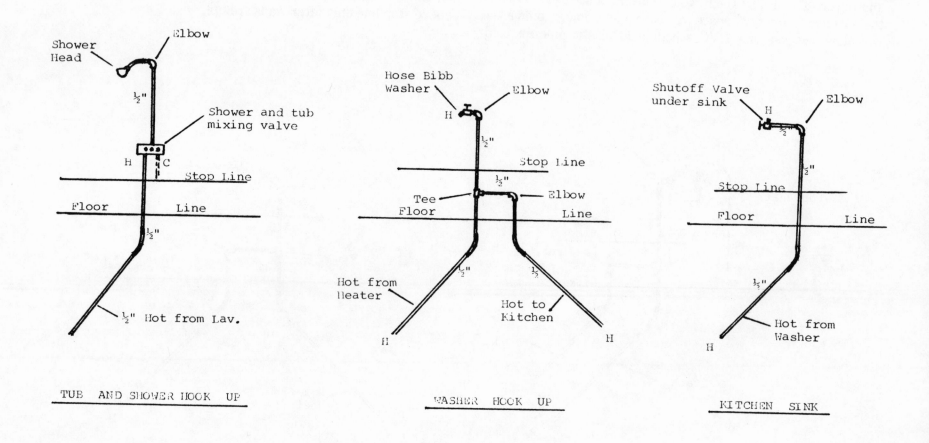

TUB AND SHOWER HOOK UP

WASHER HOOK UP

KITCHEN SINK

SEWAGE-DISPOSAL-SYSTEM SEPTIC TANK

NOTE: verify all sizes and conditions with your local codes.

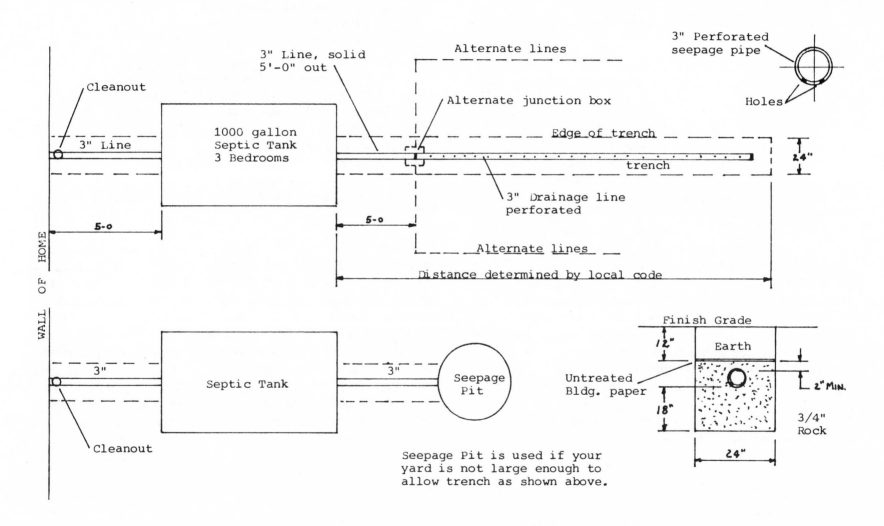

Seepage Pit is used if your yard is not large enough to allow trench as shown above.

CONSTRUCTION PROCEDURES FROM FLOOR UP

1. Snap lines for wall locations on floor.
2. Order studs, plates, headers, nails, corner braces.
3. Build all walls, complete with fire blocks.
4. Stand walls; plumb and brace securely.
5. Install trusses or rafters and brace in place.
6. Install facia board and overhang.
7. Place 2×6 drywall backing on top of parallel walls.
8. Nail on plywood roof sheathing.
9. Staple 15 lb. felt paper on roof.
10. Install all windows and outside doors and frames.
11. Celotex or wrap all outside walls completely.
12. Rough in heating-system unit, ducts, and boxes.
13. Top out plumbing waterlines and cap vents throughout.
14. Connect water pressure to lines for test.
15. Check on gas lines if available, and install.
16. Call power company for meter and panel location.
17. Install all wiring.
18. Install attic, kitchen, bath vents through roof.
19. Call for building inspection.
20. Install wall and ceiling insulation batts.
21. Call for insulation inspection, if required.
22. Hang drywall, ceilings, and all walls.
23. Call for drywall inspection, if required.
24. Tape and sand drywall throughout.
25. Spray acoustical ceilings and texture walls.
26. Install finish floor, if required.
27. Set all inside doors and frames.
28. Install exterior siding and trim.
29. Complete all inside painting of home.
30. Set all cabinets and Formica tops.
31. Set and connect all plumbing fixtures.
32. Complete wiring and heating details.
33. Install guttering, if required in your area.
34. Paint exterior of home.
35. Install floor tile in kitchen and baths.
36. Install carpet as required.
37. Grade yard and plant shrubs for landscaping.
38. Call for final inspection.

NOTE: Keep your construction site in a neat and orderly condition at all times. Many building departments will make an unexpected inspection of the site. A "red tag" from them means you must stop construction until the condition is approved.

If you follow the above steps one by one, you will find that everything will come together more quickly. Complete each step as you come to it, and you will avoid having to go back and finish up small details that could easily be overlooked. When applying for building permits, verify with your building department the proper inspection times; these may vary from area to area.

MATERIAL LIST

This list covers all materials required, from concrete slab floor up or wooden subfloor up. This list follows your footing material list, which is determined by the type of wood or concrete floor used and by the exterior siding your home requires. Refer to footing-section drawings.

Description	Quantities	Cost
2″ × 4″ × 12′ bottom wall plates		
2″ × 4″ wall studs, 92¼″ long		
1″ × 6″ × 12′ corner braces		
4″ × 6″ headers over openings		
4″ × 8″ headers		
4″ × 12″ headers		
4″ × 14″ header over garage door		
Steel-plate straps		
Double 2″ × 4″ × 12′ top wall plates		
50 lb. box 16p nails (sinkers)		
50 lb. box 8p nails		
Roof trusses (if used)		
2″ × 6″ ceiling joists (if used)		
2″ × 6″ roof rafters (if used)		
Ridge board and blocks (if used)		
Fire blocks, 2″ × 4″ × 14½″ long		
Facia board, 2″ × 6″ standard		
⅜″ plywood to box in overhang		
Plywood roof (½″ × 4′ × 8′ plywood)		
Vents and roof jacks		
Attic and overhang vents		
15 lb. building felt (rolls)		
Finish roofing shingles, etc.		
Exterior doors and frames		
Windows: aluminum or wood frame		
Window and door flashing		
Exterior wall sheathing		
Rough-in wiring		
Rough-in heating and air conditioning		
Stub-out plumbing in walls		
Installation of septic-tank system		
Insulation: walls and ceilings		
Drywall, 4′ × 8′ × ½″ sheets		
Exterior siding: lap, stucco, etc.		
Drywall tape and joint mud		
Acoustical-spray ceilings		
Spray-texture walls		
Finish floors (if wood subfloor)		
Interior doors and frames		
Window trim, baseboard, etc.		
Interior paint		
Kitchen and bath cabinets		
Counter tops: Formica, tile, etc.		
Plumbing fixtures and hookup		
Light fixtures: plugs, switches, etc.		
Floor tile: kitchen, baths, etc.		
Carpet and padding		
Overhang guttering		
Exterior painting		
Finish grading		
Landscaping trees and shrubs		
TOTAL COST		

Refer to "Material List Breakdown" for explanation of each item.

WALL-FRAMING SCHEDULE

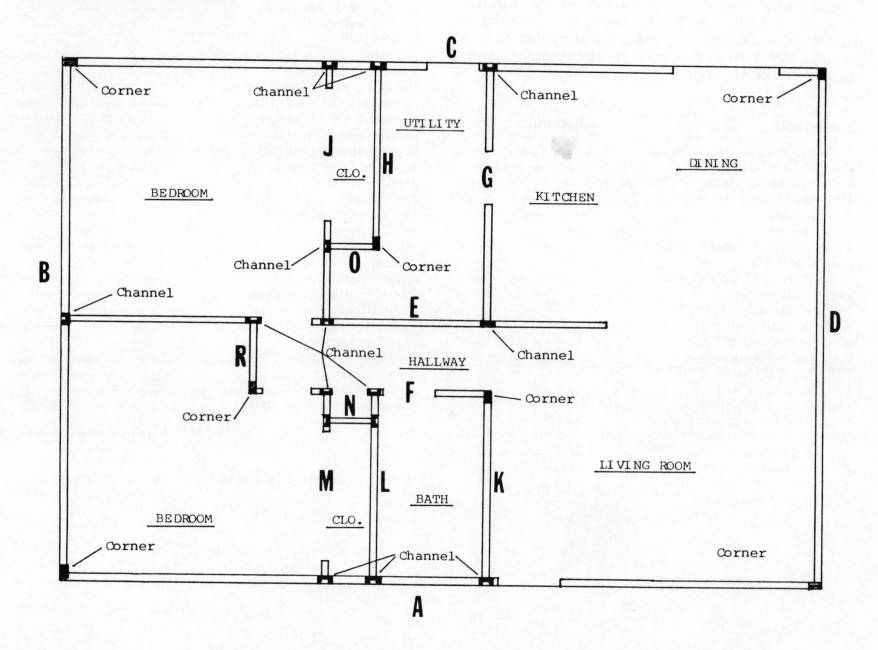

ROOM LAYOUT ON SUBFLOOR OR CONCRETE SLAB

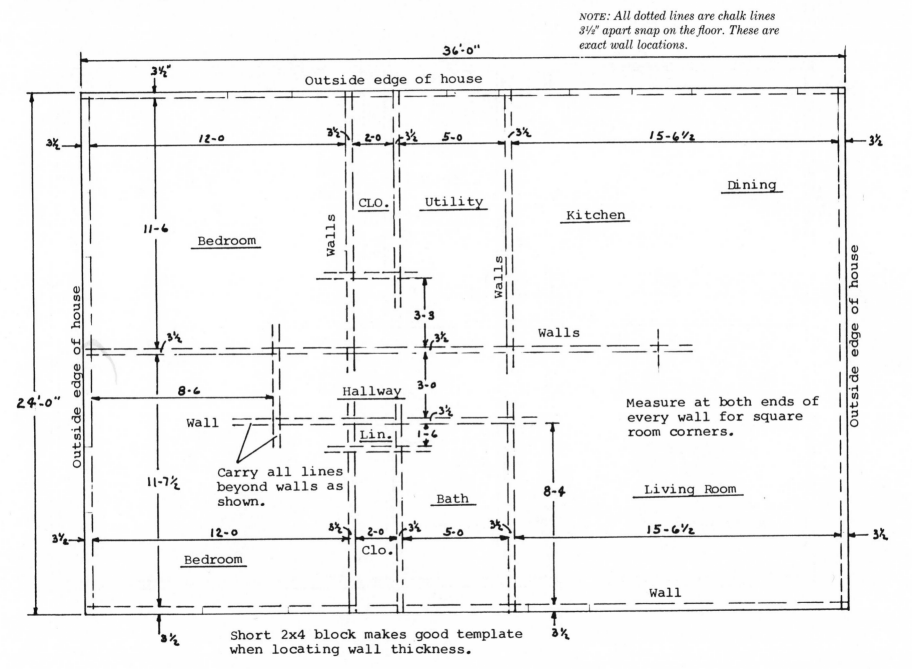

NOTE: All dotted lines are chalk lines 3½" apart snap on the floor. These are exact wall locations.

Outside edge of house

36'-0"

3½"

12-0   3½   2-0   3½   5-0   3½   15-6½   3½

Dining

CLO.   Utility   Kitchen

Bedroom

Walls   Walls

11-6

24'-0"

Outside edge of house   Outside edge of house

3½

Walls

3-3

3½

8-6   3-0

Hallway

Wall

Carry all lines beyond walls as shown.

Lin.   3½

1-6

Measure at both ends of every wall for square room corners.

11-7½   8-4   Living Room

Bath

12-0   3½   2-0   3½   5-0   3½   15-6½   3½

Bedroom   Clo.

Wall

3½   3½

Short 2x4 block makes good template when locating wall thickness.

TYPICAL DOOR AND WINDOW SIZES

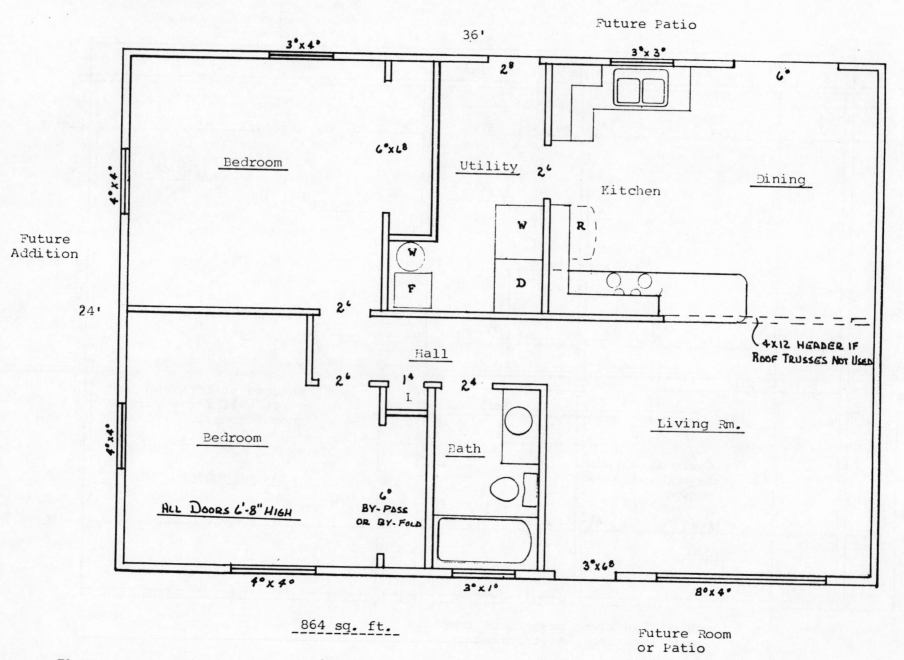

Future Patio

36'

3⁰ x 4⁰

3⁰ x 3⁰

2⁸

6⁰

Bedroom

6⁰ x 6⁸

Utility

2⁶

Kitchen

Dining

4⁰ x 4⁰

W

R

Future
Addition

W

F

D

Futur
Garag

24'

2⁶

4 x 12 HEADER IF
ROOF TRUSSES NOT USED

Hall

2⁶

1⁴

2⁴

I

Living Rm.

Bedroom

4⁰ x 4⁰

Bath

ALL DOORS 6'-8" HIGH

6⁰
BY-PASS
OR BY-FOLD

3⁰ x 6⁸

4⁰ x 4⁰

3⁰ x 1⁰

8⁰ x 4⁰

864 sq. ft.

Future Room
or Patio

ROUGH FRAMING FOR DOORS AND WINDOWS

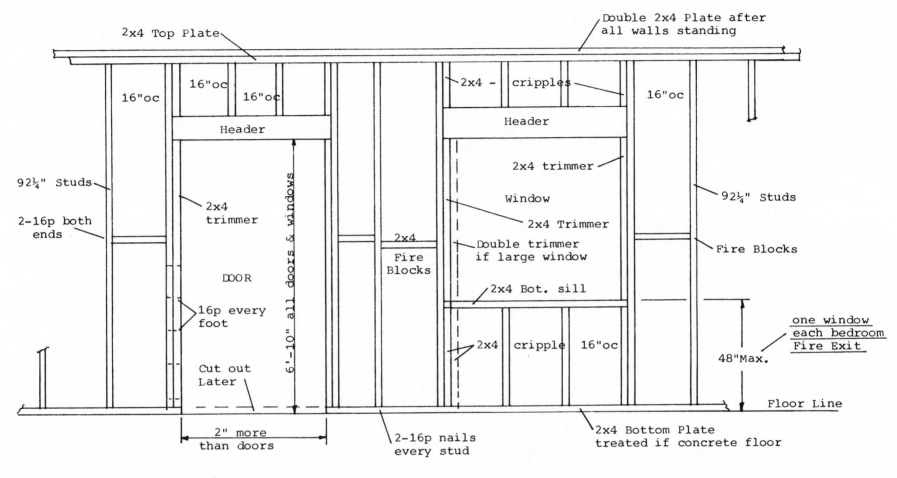

Window supplier will furnish
rough opening dimensions.

Typical Header Sizes:
0' to 4' = 4x6
4' to 6' = 4x8
6' to 8' = 4x12
Garage Door 4x14

TYPICAL WALL FRAMING

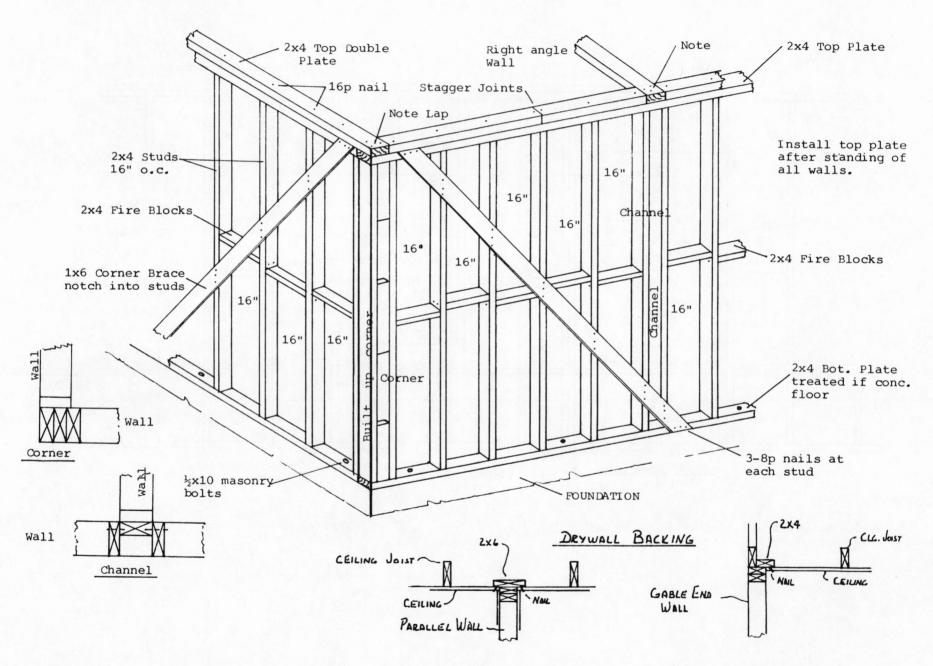

72

TOP-WALL-PLATE INSTALLATION DETAILS

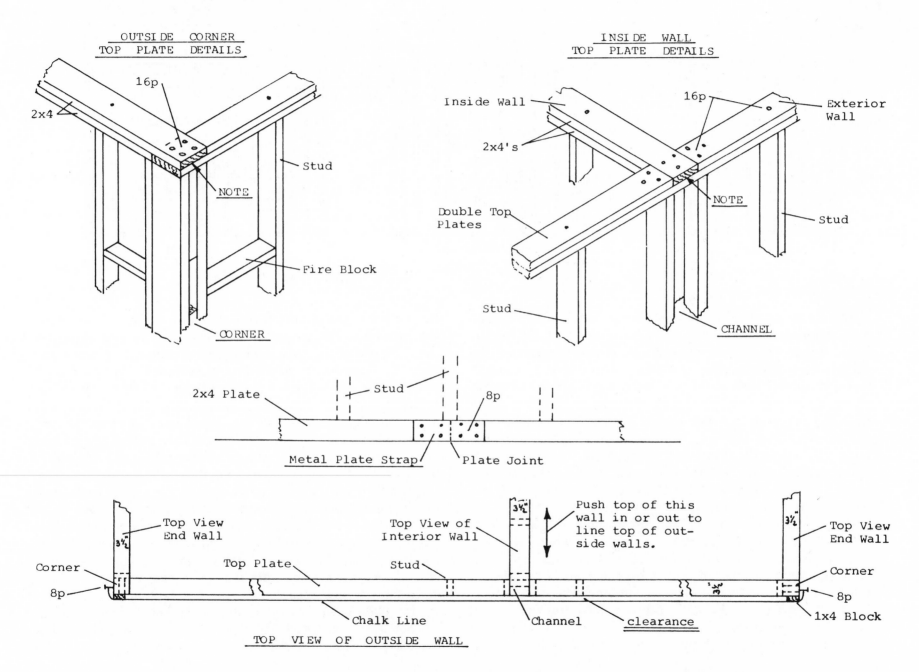

OUTSIDE CORNER
TOP PLATE DETAILS

16p

2x4

NOTE

Stud

Fire Block

CORNER

INSIDE WALL
TOP PLATE DETAILS

Inside Wall

16p

Exterior Wall

2x4's

Double Top Plates

NOTE

Stud

Stud

CHANNEL

2x4 Plate

Stud

8p

Metal Plate Strap

Plate Joint

Top View End Wall

Top View of Interior Wall

Push top of this wall in or out to line top of outside walls.

Corner

Top Plate

Stud

Corner

8p

8p

Chalk Line

Channel

clearance

1x4 Block

TOP VIEW OF OUTSIDE WALL

FRAMING OF WALL "A"

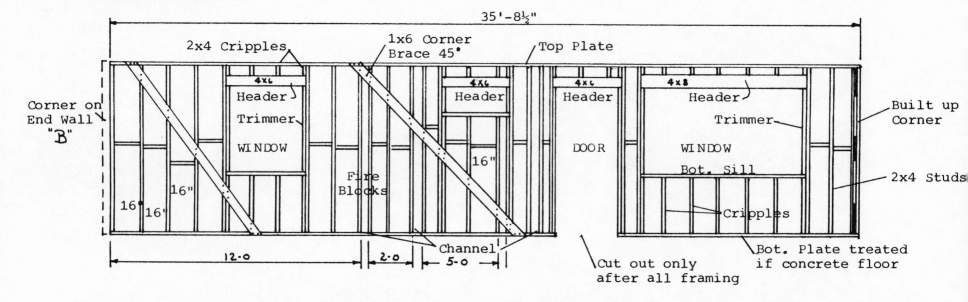

You are now standing on your subfloor, concrete or wood. You have snapped chalk lines 3½″ apart, locating every wall in your home. First, lay bottom plate 2×4's completely around the outside edge of your home or subfloor. These should lie end to end. Cut these 2×4's at the corners as needed to lay them flat on the floor. Next, lay bottom plate 2×4's on all interior-wall lines and cut them to match the chalk lines. You now have all the bottom plates laid out. If you are working on a concrete floor, you will, with square and pencil, have to mark the bolt holes on the outside bottom-plate walls. Also, be sure to measure in from the subfloor edge to the center of every bolt and locate them on the bottom plate. Next, drill the ¾″ holes in the bottom plate for each bolt. If you are working on a wood floor, omit the above drilling since no bolts are showing. Next, lay the top-plate 2×4's on top of the bottom plates and cut to exactly the same length, but do not have any joints together in the top and bottom plates. These joints must be staggered at least 48″ apart along the wall. You now have top and bottom plates cut for the length of each wall. Next, holding these plates together, turn both up on edge to mark each stud loca-

tion on the edges of both plates at the same time. First, mark the location of the corner, as shown in the drawing above, at the right end of the wall; from the very end of the plates measure 16″ to the center of the first stud from the corner. Next, continue the full length of plates, marking studs every 16″ on center to the end. Noting the chalk lines on the floor, you will mark the edge of the plates for each channel; a channel is installed where any wall joins another wall at a right angle. You must be accurate. A 2×4 channel allows you to nail drywall in the corners of a room. From your floor plan, locate the door and windows on the edge of the plates. You must have your rough window-opening sizes to frame these openings. Your window dealer will furnish these measurements. Study the detail drawings on how to frame an opening. Many builders will build all channels, corners, and window and door openings first. Then, when they start building walls, everything goes much faster. Now separate the top and bottom plates, laying the top plate in toward the center of the house. Plates will lie on edge 92¼″ apart; this will allow a stud to be placed on edge at each stud location as marked on the plates.

FRAMING OF WALL "C"

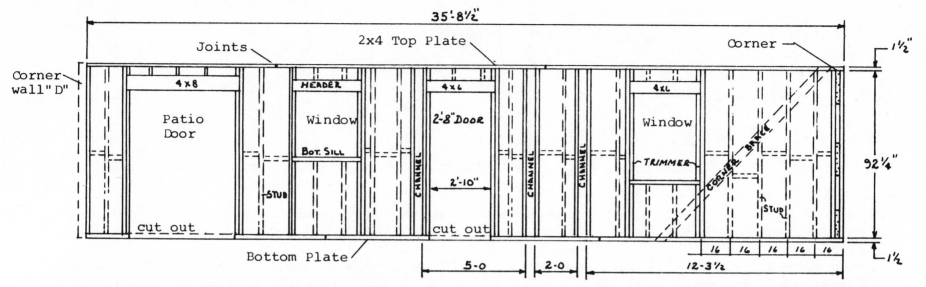

You now have the top and bottom plates lying on edge, on the subfloor. With these spaced 92¼" apart, next lay a stud on edge at each mark on the plates. Put the corner assembly in place at the end of the wall; then set all window and door assemblies in place according to your marks on the plates. When laying out stud locations on the top and bottom plates, it will help to mark both plates with a "C" for channels, a "W" for windows, and a "D" for doors. Next, with all doors, window assemblies, channels, and corners in exact position, nail these assemblies to the top and bottom plates of the wall. Use two 16p nails and nail through the plate and into the end of every 2×4 in these assemblies, top and bottom. Be sure every member is in its proper place before nailing. Double-check your dimensions. Next, lay a stud on edge at each mark on the plates and nail these in place: two 16p nails through the plates into the end of the stud at both ends. Do not short-nail a wall. You still need short 2×4's above and below all the window openings. Cripples (see Glossary) are cut to fit at each mark. If you have short pieces of 2×4 above the windows, toenail these in place with 8p nails to avoid splitting them. Did you place the treated bottom plate on the bottom of the wall, if your floors

are concrete? You now have every 2×4 nailed into the wall. Now, with a 50' steel tape, measure from the top of the left end of the wall to the bottom of the right end of the wall and note this dimension. Next, measure from the bottom of the left end of the wall to the top of the right end of the wall. This dimension must be exactly the same as the previous measurement. If it is not, then kick the bottom plate of the wall to the left or right and remeasure. Kick and measure until both measurements are the same. You now have a square wall, which is a must. Next, lay your 1×6 corner brace on a 45-degree angle from the top to the bottom of the wall. (See drawings and note that the top of the corner brace always points toward an outside wall.) With the corner brace in place, mark a line at both edges of the 1×6 on the edge of every stud that a corner brace covers, including top and bottom plates. Set your power saw to cut both lines ¾" deep on every stud and plate covered by the 1×6 corner brace. Using a 1"-wide wood chisel and a hammer, split out each notch ¾" deep in every stud. This will allow the 1×6 to lie flush in the notch of each stud. At the bottom plate, nail the 1×6 into the notch with three 8p nails; then cut off the top and bottom of the 1×6 flush with the edge of the plates.

FRAMING OF WALL "D"

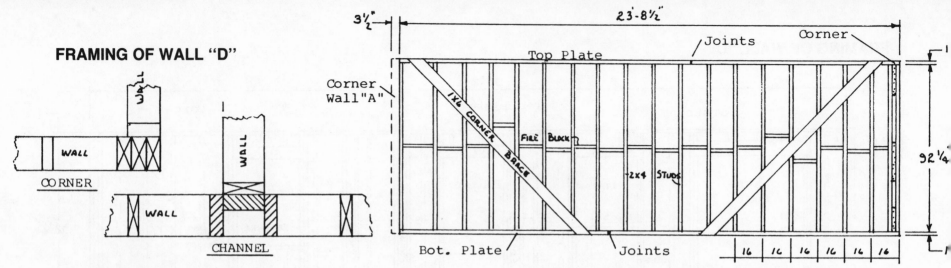

With 8p nails, tack the 1×6 corner brace into the notches in three different studs up the wall. Do not drive these nails all the way in; they must be pulled later when you stand the walls. You are now ready to install fire blocks; these are very important—mainly for strength in the walls. Snap a chalk line down the center of the height of the wall for the fire-block location. Measure between every stud for the exact length of each block; do this near the top or bottom plate. Some studs may be warped, and if you cut the blocks to proper length they will help straighten a warped stud when nailed in place. Example: If your studs are on exact 16″ centers you would cut your block 14½″ long. Most of your fire blocks will be 14½″ long if you have held your 16″ centers. Some blocks will be shorter; use every piece of scrap 2×4 on the place for fire blocks before cutting up good 2×4's. Refer to detail drawings; notice how fire blocks are nailed in place, staggered 1½″; this will allow you to put two 16p nails through the stud and into the end of every block, both ends. You have now completed a true and square wall ready for standing. But do not stand it yet; do another wall—"B" or "D". Wall "D" will go up quickly with no windows or doors.

Again, lay your top and bottom plates 92¼″ apart on the floor. Place studs on edge at every mark on plates and nail them to the plates. Next do your corner braces, then the fire blocks; another wall is built. Remember, nail your corner brace securely at the bottom plate only, tack it the rest of the way up the wall. Before moving a wall out of the way, take a metal plate strap 1½″ wide and 8″ long with nail holes in it, purchased from your lumber yard. Nail this plate strap on the edge of the 2×4 top and bottom plates at every joint of 2×4 plates, making sure the 2×4 plates are tight together before nailing. Nail a strap on the other side after the wall is standing in place. Next, you should build some interior walls such as walls "K" and "G"; you will need these interior walls to serve as props when you stand some of the exterior walls. Again, set the top and bottom plates on edge 92¼″ apart; lay in studs and nail. Measure the corners both ways for a square wall, notch in the corner braces and tack in place; add fire blocks and another wall is ready. Remember to make the corner to corner measurement on every wall before marking the corner-brace location. You must make sure your walls are square; you will be in trouble later when you're standing walls if the walls are not square. Next, build all your short walls such as "R," "N," "O," "H"; stack these out of the way. When standing walls, you can grab one of these quickly and nail it in place on the chalk lines and use it to tie or brace a long heavy wall. If you are working on a concrete floor, it will be necessary to rent a .22-caliber blank gun that shoots a large nail down through a steel plate (a bottom 2×4 plate) and into the concrete slab to hold interior walls in place. If you are working with a wood floor, use 16p nails every 12″ down through the bottom plate and into the double floor joist under the subfloor.

FRAMING OF WALL "B"

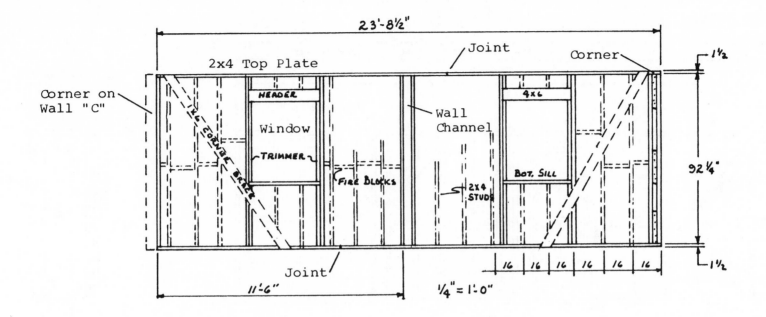

NOTE: If you find that you have nailed a channel into the wall backward (and you probably will), do not take the wall apart to change it. Using a strip of ½″ scrap plywood about 3″ wide, tack this inside the channel. Next, stand another 2×4 stud in the channel and nail it in place. This will make four 2×4's in the channel instead of the normal three. This is the quickest and easiest way to solve this problem.

You are now about ready to stand the walls. When you do this, you will be able to see what your home is going to look like. Before standing the walls, take the wood or steel stakes you used for your foundation and remove the used nails. Around all the outside walls, drive a stake securely into the ground about 6′ out from the outside wall of the house and about every 8′ around the outside of the house. At each stake lay a 10- or 12-foot 2×4 as a brace to hold the walls up and in place. You are now ready to stand the walls. Refer to page 81, Standing and Plumbing of Walls.

FRAMING OF WALL "E"

FRAMING OF WALL "R"

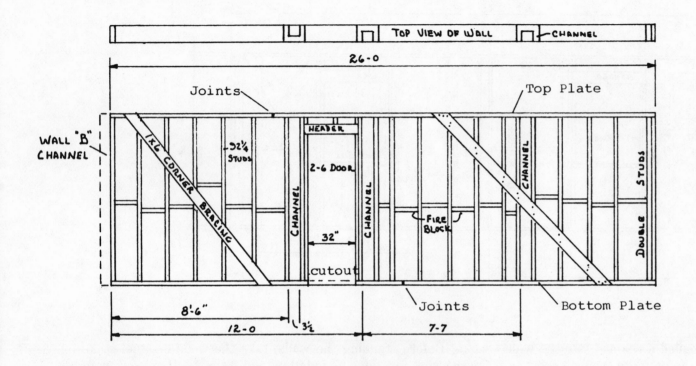

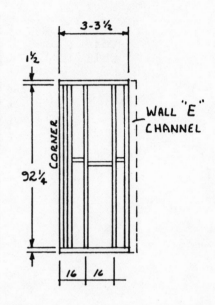

READING OUR TAPE MEASURE

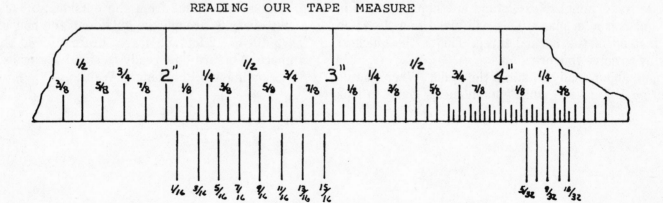

WALL-FRAMING DETAILS

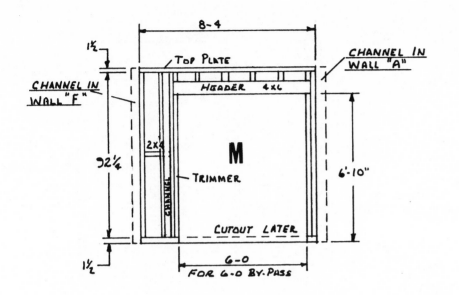

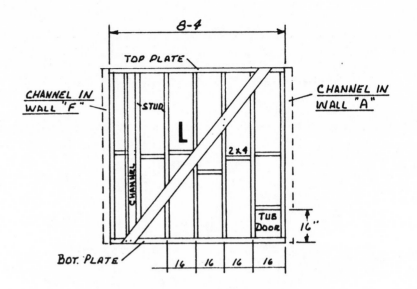

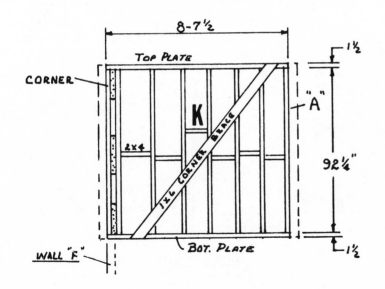

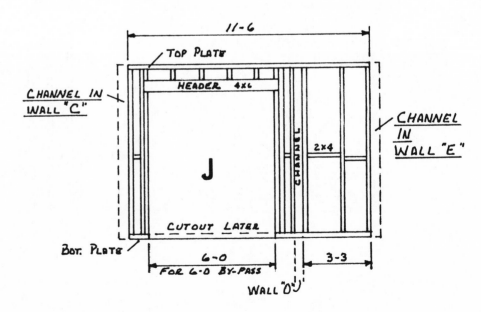

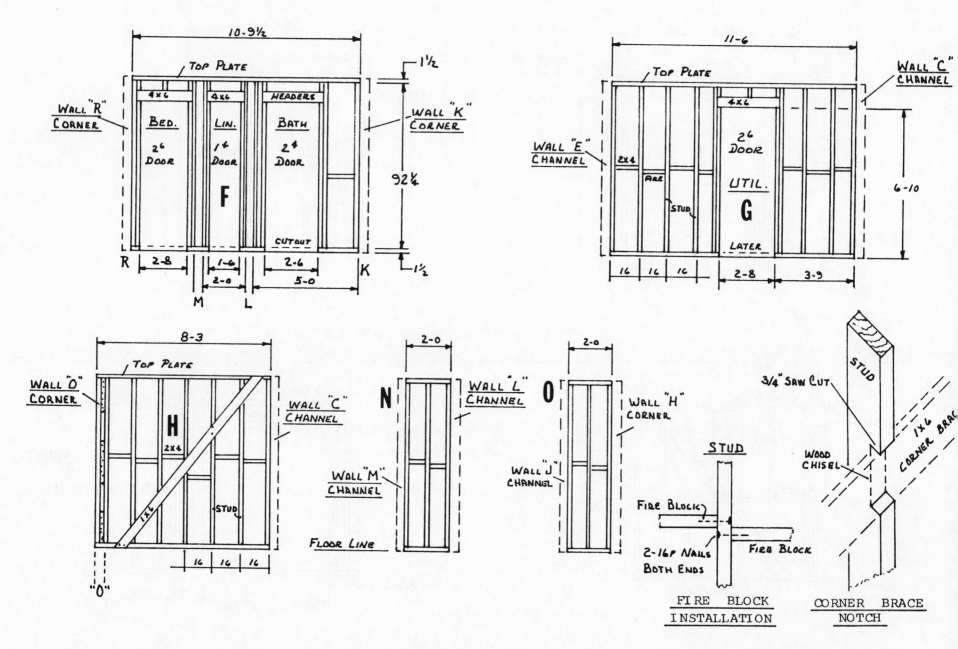

STANDING AND PLUMBING OF WALLS

Stand the front wall of your home first. This will require the assistance of three or four ablebodied friends and relatives. You must be very careful when standing a long, heavy wall. Do not let this wall fall over on someone. Lay this wall in place on the subfloor with the bottom plate as near to the correct position as possible. If you are on a concrete floor, line up the holes in the bottom plate with the bolts in the floor slab. (Forget the bolts if you are on a wood subfloor.)

With all the people spaced equally along the top plate of the wall, pick up the top of the wall slowly and raise it all the way up to standing position.

Next, starting at one end of the wall, lift a portion of the wall and set it down over the bolts. If there are no bolts, set the wall *exactly* on the chalk lines. Next, drive a 16p nail down through the bottom plate and subfloor and into the box end plate. Do this about every 6' along the wall.

Everyone is still holding this wall up. Have one man now go outside to the stakes and place a 2×4 brace from a stake up on an angle toward the top of the wall. Nail the 2×4 brace to the stake. At the top of each brace, before nailing it to a stud in the wall, place your level on the wall to be sure that the wall is standing up on perfect vertical plumb. Now nail the brace to the side of the stud in the wall. Nail it securely with a 16p nail to hold the wall in place. Remember, this wall could fall into the house also.

Next, stand wall "L" in place between the chalk lines and nail the bottom plate to the floor. Line up the end of wall "L" with the channel standing in the front wall and nail the wall to the channel with a 16p nail every foot, top to bottom. Wall "L" will keep the front wall in place.

Next, set an end wall in the same manner as the front wall. First set it in place, then make sure it's straight up and down, then add braces. Follow this wall with another inside wall. However, do not get yourself in a position where a wall cannot be moved into place because other walls have already been installed. Example: If you have installed walls "J," "G," and "E," you could not move wall "H" into place. Remember, always

anchor the bottom plate of a wall first, then line up the corners and nail them, followed by an inside wall to be nailed to channels. CONGRATULATIONS ARE IN ORDER: You now have all the walls standing in place with bottom plates anchored in place. All corners and channels are properly nailed. Leave all the braces around the outside of the house until later. Do not remove them at this time.

You are now ready to plumb and line all the walls. Start with the front wall. Remember the 1×6 corner braces you notched into this wall? You nailed this 1×6 securely to the bottom plate and just tacked it on up to the top. Now pull out the 8p nails holding the 1×6 corner brace to the stud wall. Remove these nails from all the corner braces in this wall.

Next, place your level up and down on the end or corner of the front wall and check for level. By pushing the top of this front wall toward one end of the house or the other, you will be able to vertically plumb or level both front corners of your home. Once these corners are plumb, nail your 1×6 corner brace to each stud, the full height of wall, using three 8p nails at each stud. You may need to use a long 2×4 as a brace from the ground up to the top of the wall for leverage at one end of the wall. Do not nail the corner braces until they are on a perfect plumb.

Nail both ends of the front wall to the corners of the two end walls. Now this end of both end walls is plumb with the front wall. Continue around the outside walls and plumb every corner both ways and nail the corner braces as each wall is plumbed.

Remove the 8p nails from the corner brace of all the interior walls anchored to the floor and end channels. Use a ladder to climb to the top outside corners of all outside walls. Tack a small scrap of 1×4 at both top outside corners of each wall.

Next, pull a chalk line, the full length of the wall, tight against the scrap 1×4 at each end of the wall. This chalk line will show you if the top of the wall is bowed in or out. If the top of wall is bowed in, push the top of the interior wall out until the space between the chalk line and the wall is the same the full length of wall. When the top of the wall is straight, nail the corner brace into the interior wall to keep all walls plumb.

Follow this same procedure on all walls. With the top of all walls straight and leveled in all directions, and all channels and corners nailed with 16p nails, you are ready to install the double top plate. This is the second 2×4 on top of all wall top plates.

First, install this 2×4 on top of all interior walls. This plate must extend out over the top of the exterior walls and be flush with the outside edge of the house. This helps tie the tops of all the walls together. Do this also wherever any interior wall ties into another interior wall.

Refer to detail drawings (page 79) showing the lap of the top plates at the corners and interior walls. The next step will be Ceiling Drywall Backing, Ceiling Joist and Roof Framing.

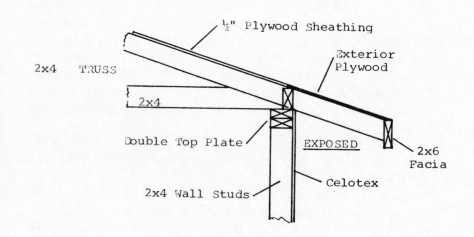

ROOF OVERHANG DETAILS

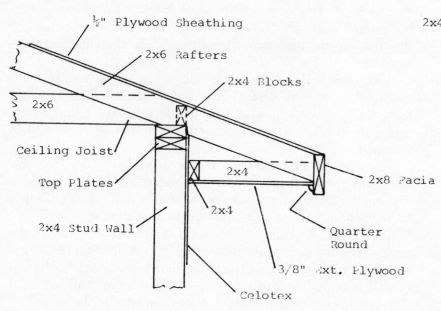

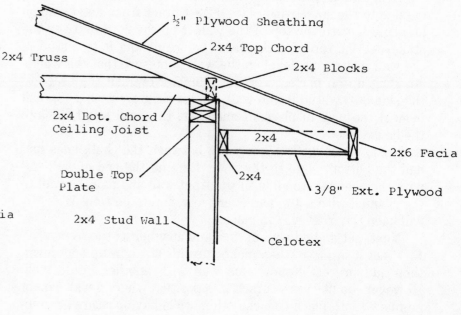

82

CEILING JOIST LAYOUT IF ROOF TRUSSES NOT USED

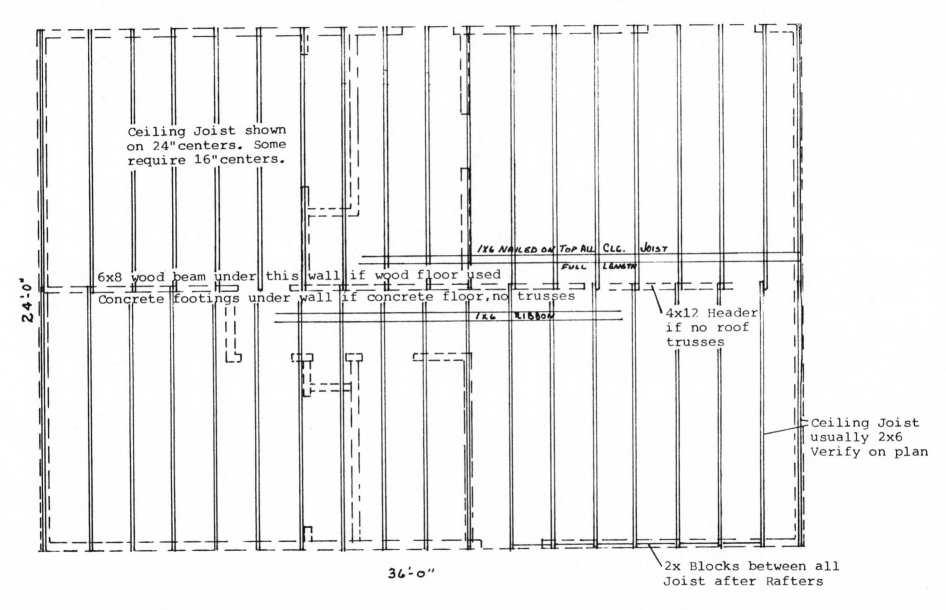

Ceiling Joist shown on 24"centers. Some require 16"centers.

24'-0"

6x8 wood beam under this wall if wood floor used

Concrete footings under wall if concrete floor, no trusses

1X6 NAILED ON TOP ALL CLG. JOIST

FULL LENGTH

1X6 RIBBON

4x12 Header if no roof trusses

Ceiling Joist usually 2x6 Verify on plan

36'-0"

2x Blocks between all Joist after Rafters

NOTE: *Roof trusses will save money and time.*

STANDARD FACTORY BUILT TRUSS

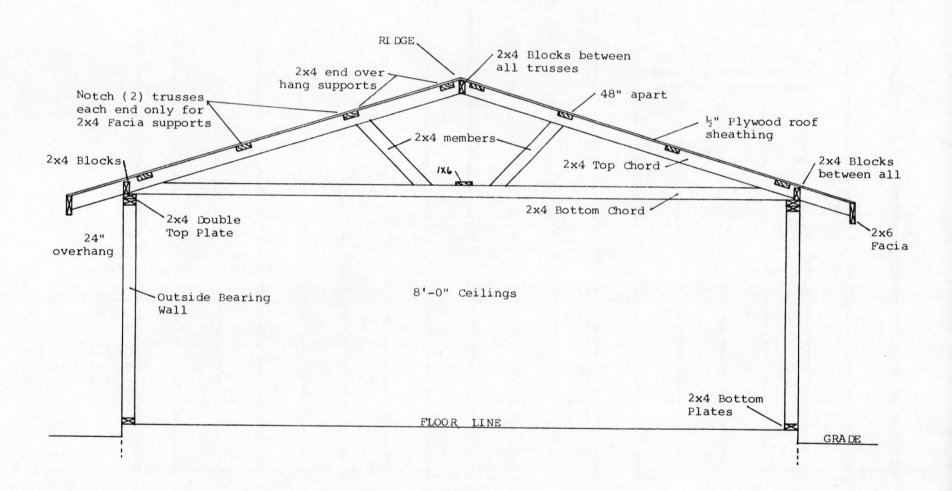

RIDGE

2x4 Blocks between all trusses

2x4 end over hang supports

Notch (2) trusses each end only for 2x4 Facia supports

48" apart

2x4 members

½" Plywood roof sheathing

2x4 Blocks

1X6

2x4 Top Chord

2x4 Blocks between all

2x4 Double Top Plate

2x4 Bottom Chord

2x6 Facia

24" overhang

Outside Bearing Wall

8'-0" Ceilings

2x4 Bottom Plates

FLOOR LINE

GRADE

84

STANDARD FACTORY GABLE END TRUSS

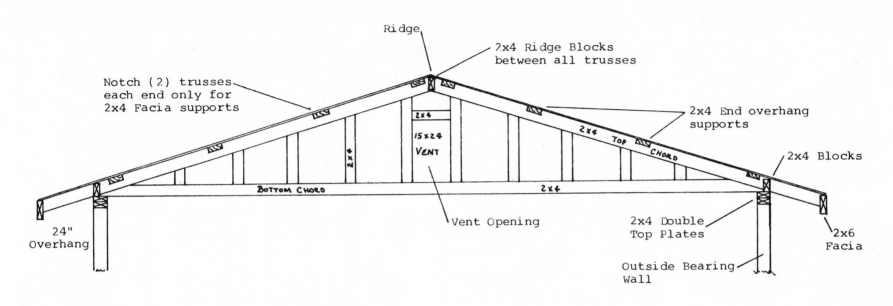

Ridge

2x4 Ridge Blocks
between all trusses

Notch (2) trusses
each end only for
2x4 Facia supports

2x4 End overhang
supports

2x4 Top Chord

2x4

15x24 Vent

2x4 Blocks

2x4 Bottom Chord

Bottom Chord

2x4

Vent Opening

2x4 Double
Top Plates

2x6 Facia

24"
Overhang

Outside Bearing
Wall

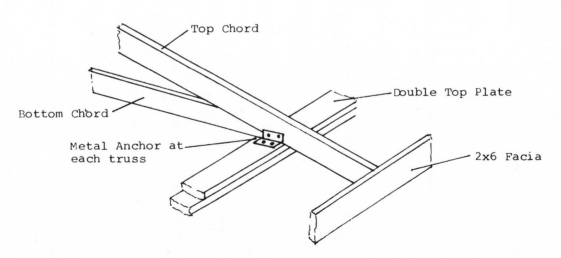

Top Chord

Double Top Plate

Bottom Chord

Metal Anchor at
each truss

2x6 Facia

ROOF TRUSS FRAMING LAYOUT

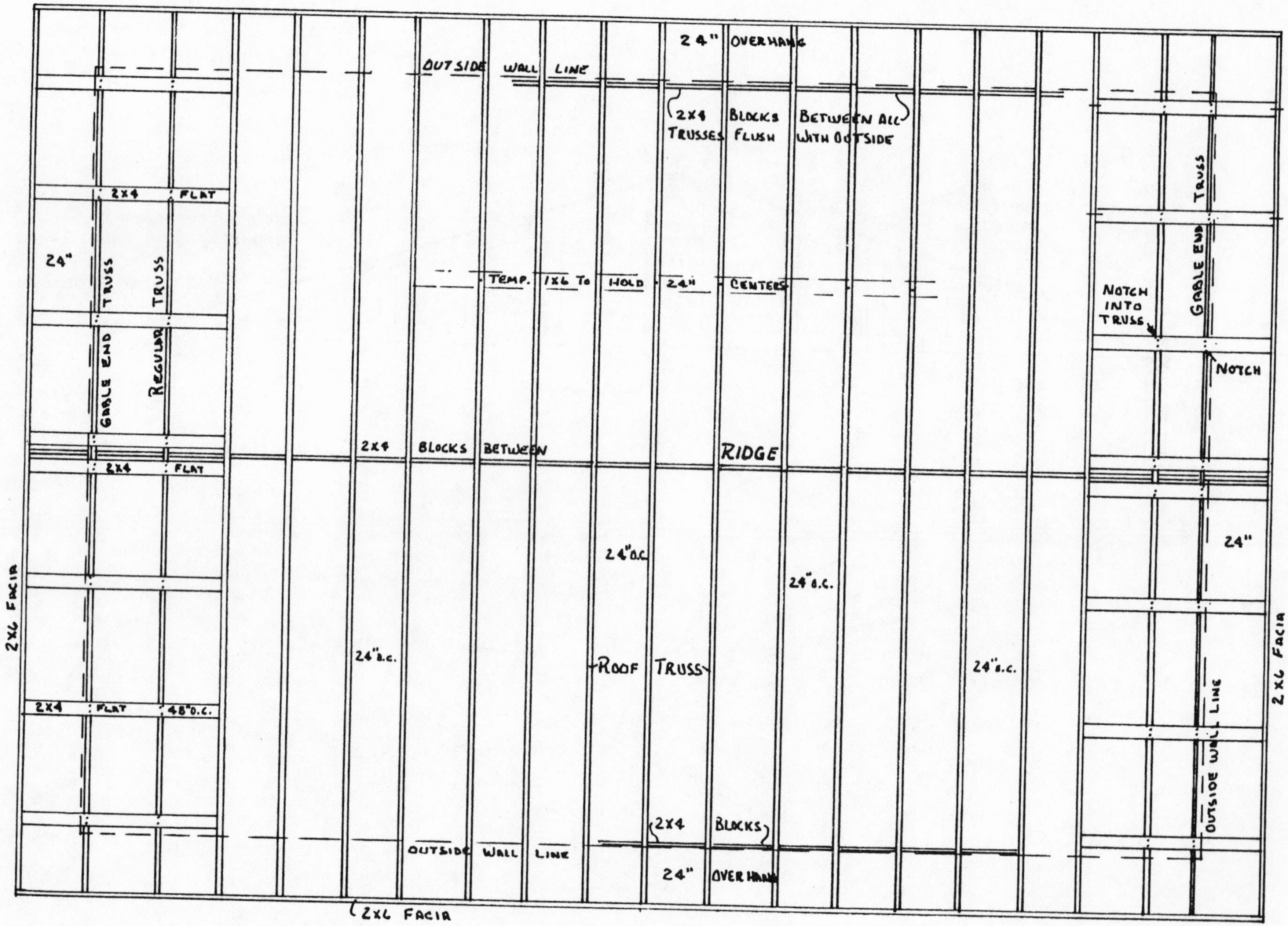

24" OVERHANG

OUTSIDE WALL LINE

2X4 BLOCKS BETWEEN ALL TRUSSES FLUSH WITH OUTSIDE

2X4 FLAT

GABLE END TRUSS

24"

REGULAR TRUSS

GABLE END TRUSS

NOTCH INTO TRUSS

TEMP. 1X6 TO HOLD 24" CENTERS

NOTCH

2X4 BLOCKS BETWEEN

RIDGE

2X4 FLAT

24"

24"O.C.

24" O.C.

2X6 FACIA

24"O.C.

24" O.C.

ROOF TRUSS

24" O.C.

2X4 FLAT 48"O.C.

OUTSIDE WALL LINE

2X6 FACIA

2X4 BLOCKS

OUTSIDE WALL LINE

24" OVERHANG

2X6 FACIA

ROOF RAFTER FRAMING

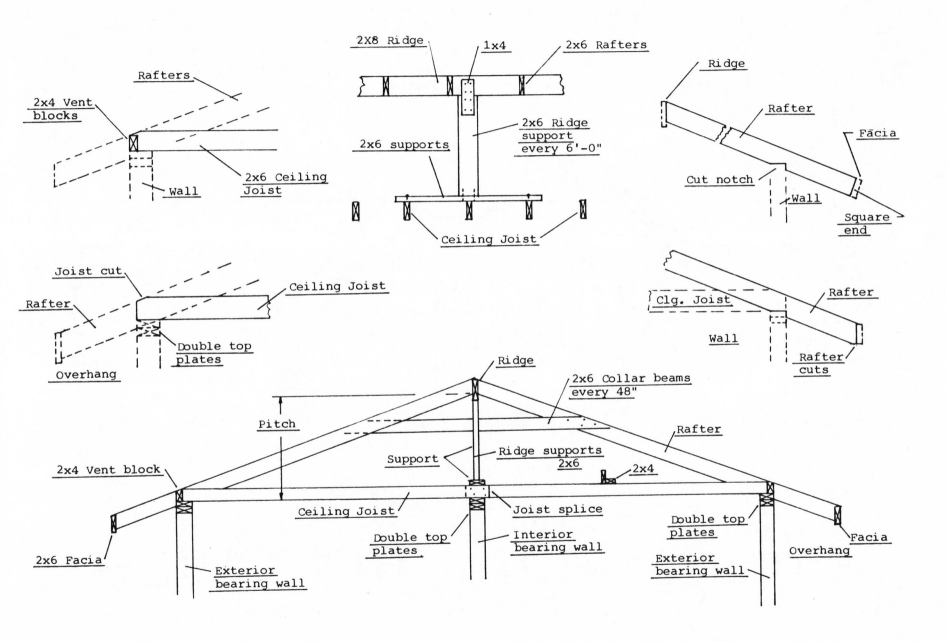

HIP ROOF FRAMING LAYOUT

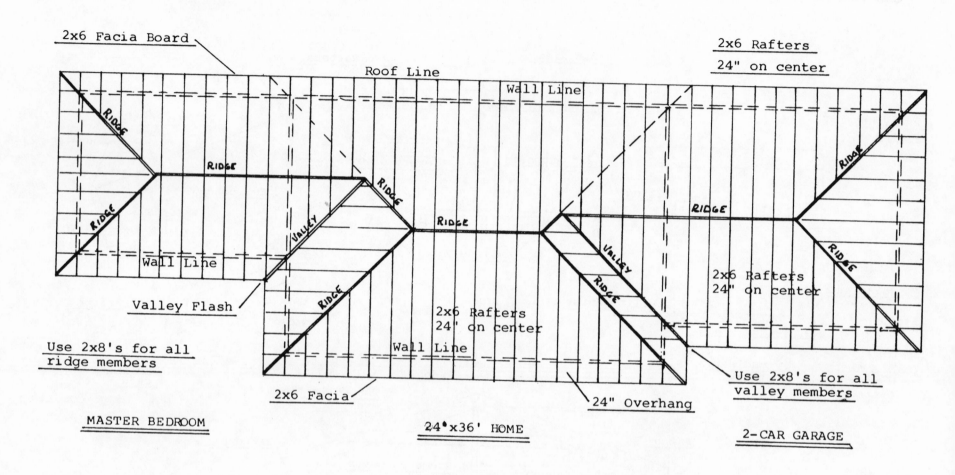

2x6 Facia Board

Roof Line

2x6 Rafters
24" on center

RIDGE

RIDGE

RIDGE

Wall Line

VALLEY

RIDGE

RIDGE

RIDGE

Wall Line

Valley Flash

Use 2x8's for all
ridge members

2x6 Rafters
24' on center

Wall Line

2x6 Facia

MASTER BEDROOM

VALLEY

RIDGE

2x6 Rafters
24" on center

24" Overhang

Use 2x8's for all
valley members

24'x36' HOME

2-CAR GARAGE

Scale: 1/8"=1'-0"

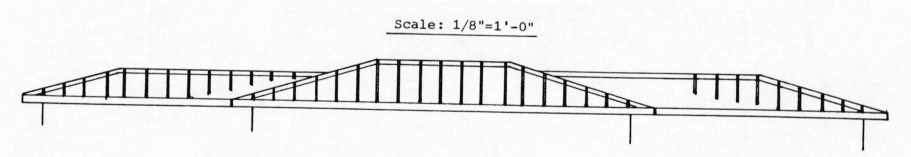

PLYWOOD ROOF SHEATHING LAYOUT

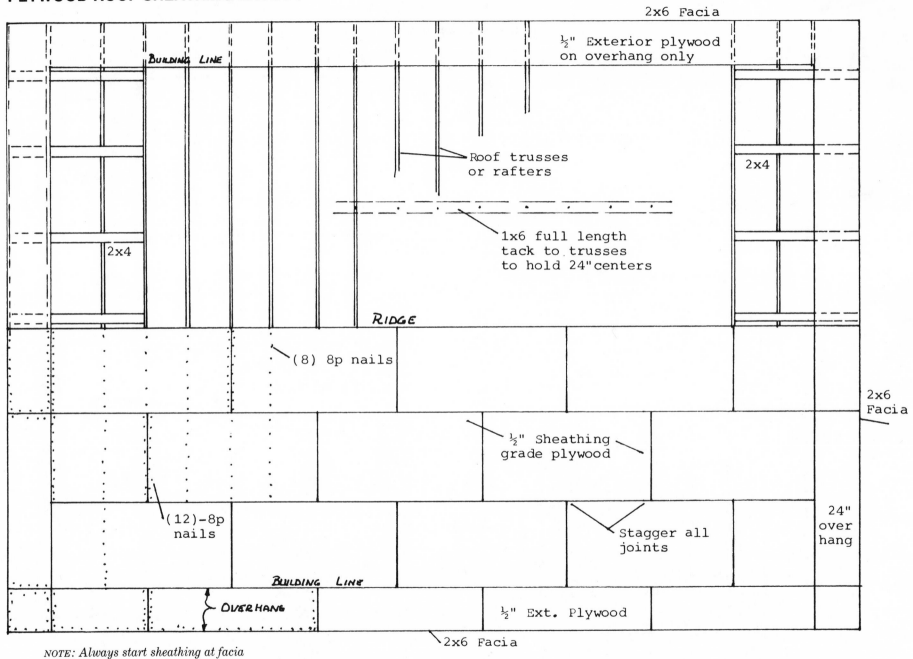

2x6 Facia

½" Exterior plywood on overhang only

BUILDING LINE

2x4

Roof trusses or rafters

1x6 full length tack to trusses to hold 24"centers

2x4

RIDGE

(8) 8p nails

2x6 Facia

½" Sheathing grade plywood

(12)-8p nails

Stagger all joints

24" over hang

BUILDING LINE

OVERHANG

½" Ext. Plywood

2x6 Facia

NOTE: Always start sheathing at facia and work uphill.

PATIO CONSTRUCTION DETAILS

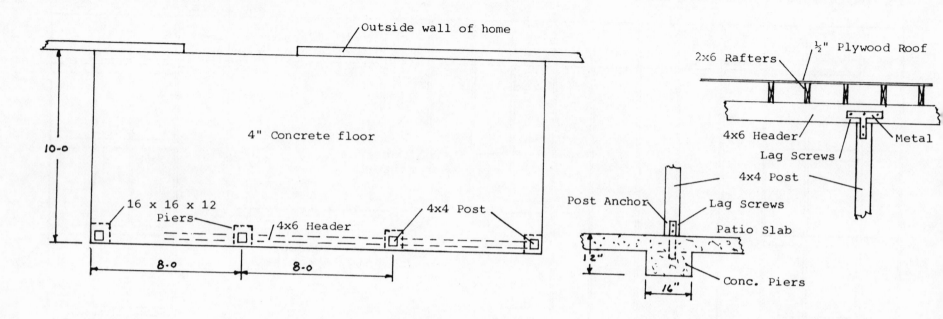

Outside wall of home

4" Concrete floor

10-0

16 x 16 x 12 Piers

4x6 Header

4x4 Post

8.0   8.0

2x6 Rafters

½" Plywood Roof

4x6 Header

Metal

Lag Screws

4x4 Post

Post Anchor

Lag Screws

Patio Slab

12"

16"

Conc. Piers

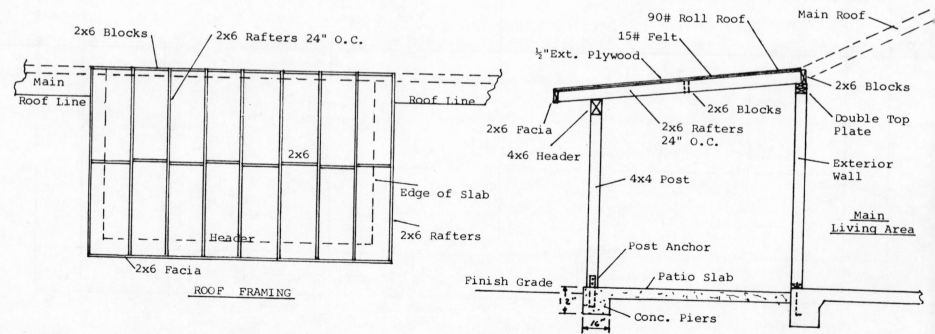

2x6 Blocks

2x6 Rafters 24" O.C.

Main Roof Line

Roof Line

2x6

Edge of Slab

2x6 Rafters

Header

2x6 Facia

ROOF FRAMING

90# Roll Roof

Main Roof

15# Felt

½"Ext. Plywood

2x6 Blocks

Double Top Plate

2x6 Facia

2x6 Blocks

4x6 Header

2x6 Rafters 24" O.C.

Exterior Wall

4x4 Post

Main Living Area

Post Anchor

Patio Slab

Finish Grade

12"

16"

Conc. Piers

90

FACTORY-BUILT METAL FIREPLACE

This is the time to install a fireplace if you plan to have one. Shop around for the many types available, and price them all. The prices of these fireplaces are not out of reach and you will find it safe and foolproof. When you have found the right unit, the maker's literature will give you dimensions for A, B, C, and D as shown on the drawings on this page. If your home will have a concrete floor and you plan on installing the unit on the outside wall, read the following suggestions.

Even if your home has wood floors, the same slab can be installed. The supplier will provide all parts for complete assembly of this fireplace, except wood framing.

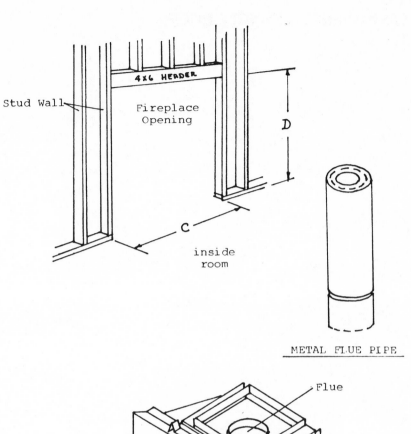

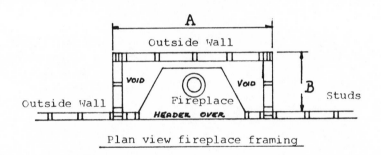

Plan view fireplace framing

Form and pour an 8″-deep concrete slab 24″ out and 60″ long at the desired location. Make the top of the slab even with the floor of the house. Set regular masonry bolts in the slab to anchor the outside walls of the fireplace cavity. Frame these walls the same as all other walls, fireblocks, treated bottom plates, etc. These walls should extend 24″ above the high point of the roof and be fully insulated before the siding is installed. Follow manufacturer's instructions.

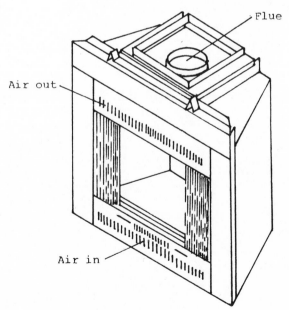

235# ASPHALT SHINGLE ROOF

NOTE: (3) Bundles of shingles covers
(1) square or (100) square feet of roof.

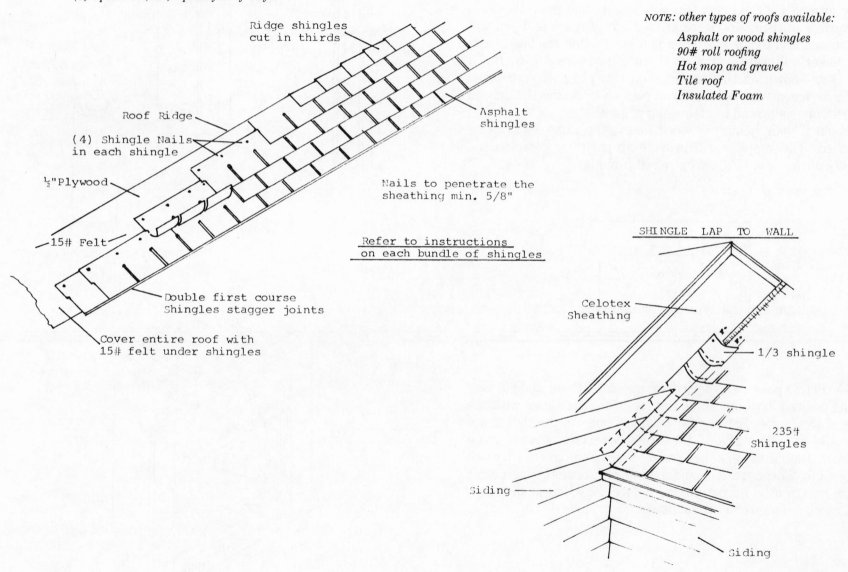

Ridge shingles
cut in thirds

NOTE: *other types of roofs available:*

Asphalt or wood shingles
90# roll roofing
Hot mop and gravel
Tile roof
Insulated Foam

Roof Ridge

Asphalt
shingles

(4) Shingle Nails
in each shingle

½"Plywood

Nails to penetrate the
sheathing min. 5/8"

15# Felt

Refer to instructions
on each bundle of shingles

Double first course
Shingles stagger joints

Cover entire roof with
15# felt under shingles

SHINGLE LAP TO WALL

Celotex
Sheathing

1/3 shingle

235#
Shingles

Siding

Siding

92

HOW TO WIRE YOUR HOME

You now have the plywood on the roof and are ready to install the electric wiring. You are going to learn another profession. Follow me through step by step and you will be amazed at how much you already know about wiring a home.

First, contact your local power company. They will advise you where you must install your meter base and circuit panel. Some areas require your electrical service from their pole to your house to be underground in conduit. If this is true in your area, they will guide you for a proper installation. (This would change nothing in these drawings except that the riser pipe, the roof jack, and the weatherhead on top would be omitted.) Let's say your electrical service will be overhead.

Next, check your drawing covering electrical entrance service (page 101), to decide on the type of meter base and circuit panel you will use. It's probably best to use a flush-mounted panel, installed between the wall studs, out of the weather.

Next, "a point to ponder": Will your home be electric, or are you planning on a gas range, gas heat, gas water heater? In that case, do you have gas available in front of your house? Will the gas company tie your home in for service?

These questions will help you decide on the amps you will need for your electric service and the size of equipment you will require. If your home will be all-electric, check with your power company; you will probably need a 200-amp circuit panel. If some gas appliances are used, you may buy a 150-amp service panel.

Shop around for prices on this material; you will save.

Now that you have purchased your meter base and circuit panel, install this box between the studs at the location decided by the power company. The center of your meter should be at about eye level. Measure up from the top of the meter base to a point 3' above the roof. This gives you the length of your 1½" riser pipe. Add 48" to the pipe length and you have the length of the three wires coming down through the pipe. You are ready to start wiring your home.

Let your building department tell you what size wire to install through the weatherhead down through the pipe and into your meter base and panel. Purchase these three wires after you have installed the meter base panel and riser pipe: two black wires of the proper length and one white wire; this white wire will be your N (neutral wire). Review the drawing of meter base.

Tools: You will need a ⅜" electric drill and a ¾" bit to drill through studs and joists, side-cutter pliers to cut wire, and a wire-stripper type pliers. Your side-cutters can be used to twist wires together. Also have handy a marking crayon, steel tape, and a hammer. Try again to buy the power to operate your electric drill from a nearby neighbor. This will be much cheaper than installing a temporary power pole. Pay him well—it is still much less money.

In most areas, the building department allows the use of Romex wire for residential wiring. Check it out. To start off, purchase two boxes of 12-2-with-ground Romex, for wiring of wall plugs, switches, and lights. Each box contains 250 feet of wire. This will go a long way in the wiring of your home. Take this book with you if necessary, and purchase your copper ground rod, ground clamp, and ground wire to come out of the bottom of the panel box, which can be installed any time. See the detail drawings.

With the floor-plan drawings of your home, go through all the rooms and mark an "X" on the floor at the location of all wall plugs. Count all your switches; count all locations of wall and ceiling lights. This will tell you how many of each box you need. Buy a couple of each extra; you will need them. If plastic boxes are allowed in your area, use them.

Purchase about 20 threaded wire lock connectors as shown on the drawing of the circuit-breaker box (page 102). To install these wire lock connectors, knock out one of the slugs in the sides or bottom of the panel box. Insert a wire lock in the hole and install a nut inside the panel box. Now you are ready to install a circuit of wall plugs. Review the drawing covering wall-plug layout.

The first circuit of wall plugs will be installed around the kitchen sink. (See page 103). Since a normal cabinet-counter height is 36″ and the backsplash is 5″ high, the bottom of the plug box will be mounted 44″ up from the floor. You now have all three boxes mounted as shown. Now run a piece of Romex from the last box back to the next one toward the main panel box. With crayon, mark each stud, so you can drill through to pull the wire through from one box to the next. Mark the easiest route, with the least drilling from one box to the next.

NOTE: When I use the term *wire*, this means Romex, which has a black, a white, and a bare copper wire inside.

Now take the end of the wire out of the box and feed this wire through the holes you just drilled through the studs. Insert the end of wire into the box and out through the front of the box about 6″. Strip this 6″ of wire to show a black, a white, and a bare copper wire for attachment to the wall plug. At the next box back, where you started pulling the wire, hold the wire to the box and cut off a length long enough to go into the back of the box and 6″ out into the room. This 6″ tail is a must on every wire; the wires must go through the back of every box and out into the room 6″, so you can connect the wires.

Next, follow the same procedure to connect box (2) and box (1). Drill the holes, pull the wire, insert the wire through the back of the box at each end of the wire. Leave a 6″ tail projecting out of each box for the connections.

You are now ready to go from box (1) to the circuit breaker in the panel box. Drill the holes, pull the wire, and leave the 6″ tail out of box (1). Knock out a slug in the side of the main panel box and install a wire lock connector. Run your wire from box (1) out to the wire connector and cut a length of wire off 36″ longer. This will allow enough wire to run up through the box or panel and connect to the circuit breaker. This circuit will require a 20-amp breaker. Install the breaker, pull 36″ of

wire up through the wire lock connector in the side or bottom of the panel box, and tighten the screws gently in the connector to hold the wire.

Next, identify the circuit by writing "Kitchen Sink" on a piece of tape and wrap the tape around the wire or circuit just outside of the main panel box. Do this as each circuit is installed.

Next, remove the white plastic cover on the 36″ of Romex inside the main panel box, split the cover down the center, and peel it off. You now have a black, a white, and a bare copper wire exposed. Push the black wire back into the box neatly, then along the side of the box, and up to the circuit breaker. Make a neat square bend in the black wire, then extend it to the breaker. Cut off any extra wire, but leave all you neatly can, in case a future wiring change is required. Strip about ½″ of insulation off the end of this black wire, loosen the screw in the end of the breaker and insert ½″ of bare wire into the hole under the screw and tighten the screw securely.

Going out to the first wall-plug box on this circuit, from the main panel, find two sets of wires projecting 6″ out of the box. Split and strip both of these sets of wires from the back of the box out to the end. You now have two black wires, two white wires, and two bare wires all about 6″ long. *Always bend the white wires to the left side of the box, and the black wires to the right side of box.*

On the back of a wall plug you will see the word "White." Strip the ends of the white wires in accordance with the strip gauge (shown on the back of all wall plugs). The white wires must always be inserted into the proper holes. Note that the silver screws are for the white-wire side, the brass screws are for the black-wire side. Strip the ends of the black wires according to the gauge, and insert the tips firmly into the holes on the black side. Take the two bare copper ground wires and twist them securely together with pliers. Slip a crimp ring up over the twist and crimp securely.

Now, cut off one wire below the ring and run the other bare wire around the ground screw at the lower left corner of the wall plug. Tighten the screw securely. Push all the wires neatly, and separately, back into the box, making room for the wall plug to fit in place. Remember, when installing all wall-plug and fixture boxes, the front faces of all the boxes must extend ½″ out in front of the studs and ceiling joists, to be flush with the face of ½″ drywall when it is installed. This is "a must."

On the wall-plug wiring instructions (see page 104), please review the detail "A" drawings for illustration.

In some cases, it will be necessary to make a wiring hookup, as shown in detail "B," when a circuit must go in two different directions. In most cases, all wall plugs in a circuit will have two sets of wires per box and will be connected as directed above and in detail "A." The last box or wall plug in a circuit, however, will have only one black, one white, and one bare wire. White on the left, silver side; black on the right, brass side; bare to the ground screw. Work on only one circuit at a time. Complete it and label it.

NOTE: If you desire to install your wall plugs at this time to avoid confusion, place a piece of duct tape over the face of each plug to keep dirt and drywall mud out. As shown on the floor plan, Wall-Plug Layout, kitchens should have two circuits to take care of various appliances. Review all wiring drawings constantly.

EXTERIOR AND BATH WALL PLUGS

On the front and rear walls of your home, you should install outside wall plugs for any special lighting you want and for any outside activities that require access to an electrical outlet. In most areas, the building codes require these wall plugs. Review this as shown on page 103, Wall-Plug Layout. These outside wall plugs will require a special circuit that also includes all bathroom plugs.

This special circuit can save lives. If you happen to use an outside plug during a rain, or if it is still damp after a rain, you could receive a very severe electrical shock. If, in addition, you are standing on wet ground, you could receive a deadly shock. In the bathroom, you may drop your electric razor in the sink or lavatory full of water or turn on the radio while in the tub.

To help prevent injury during these and other dangerous situations, this one circuit takes a special circuit breaker, sometimes called a GFI breaker. A normal breaker must heat up to a certain point before it will disconnect the power through its circuit. This sometimes takes minutes after a short develops. The GFI breaker will act at once and could save a life. It's expensive but worth it if it is ever needed.

You will now install your outside wall boxes for this GFI circuit. These boxes should be set 12″ above the floor, with the opening of the box toward the outside of the house. Remember, the face of these boxes must be installed out in front of the face of the studs the proper distance to allow for the thickness of the exterior siding, stucco, or desired exterior finished walls. The face of the plug box must come out flush with the face of the exterior. These outside boxes will use a standard wall plug; you should wire these the same as for an interior wall plug. The face plate, however, comes with a small spring-loaded cap over each plug to keep out dampness.

When installing the bathroom wall-plug box, near the lavatory cabinet, remember that a future mirror will be installed over the lav. Keep this plug out of the way today. The bottom of this box should be set 36″ above the floor to allow for top- and back-splash. With crayon in hand, be ready to mark your holes to be drilled through the studs for pulling wires to these three boxes.

Starting at the main panel box, locate the holes to be drilled to the rear wall box, then to the bathroom box, and on to the front wall box. You are at last ready to pull the wire for this circuit.

Starting at the first box out from the main panel, feed your wire through the drilled holes back to the main panel. Leave the 6″ tail of wire projecting out of this box as always and pull the wire firmly toward the main panel to take up all the slack. Knock out a slug in the side of the main panel and install a wire lock connector. Holding the wire at this connector, allow an extra 36″ of wire for the GFI breaker hookup and cut the wire. Insert the 36″ of wire up through the connector and tighten the screws gently.

Next, split the plastic cover on this wire a full 36″. You now have a black, white, and bare copper wire to connect in the main panel. At this time, snap in the GFI-type breaker into the main panel. This breaker has only a coiled white wire built in. Take this wire down alongside the box and under the row of screws near the bottom of the panel. Review the drawing of the meter base panel on page 102. Loosen one of these screws and slip the bare end of this coiled wire up in the hole under the screw.

Next, taking your other white wire, cut off the extra, strip ½″ off the end, and slip the bare end up into the same hole with the coiled wire. Tighten the screw securely. Take the bare copper wire, cut some off if necessary, and insert it also in the hole as shown on the drawing and tighten the screw. Your black wire should run along the inside of the panel to the GFI breaker. Loosen the screw in the end of the breaker, strip ½″ of the insulation off this black wire, and insert the bare end in the hole under the screw and tighten firmly. Label this circuit with tape.

Starting at the first box out from the panel, drill holes to the second box out as shown in the bathroom. Pull the wire through to this box and cut it off, leaving the normal 6″ tail of wire extending out of both boxes. Again, strip the 6″ tail of wire in both boxes and strip ½″ on each of the two black and two white wires. Insert the white wires in the proper side of the plug, then the two black ones. Twist the two bare copper wires together and install a crimp ring. Clip off one wire out of the ring and attach the other wire to the screw at the bottom of the plug.

Review detail "A" on the plug-wiring drawing. Drill holes and pull the wire to the front outside wall box. Strip the 6″ tail in this final box and attach the wall plug. Verify that all materials and workmanship comply with local codes.

NOTE: After each circuit is completed, staple every wire to a stud within 12″ of every box. Be careful not to staple *into* a wire.

LIGHT FIXTURE AND SWITCH WIRING

Refer to the Light Fixture and Switch Layout floor-plan drawing on page 105. At this time, you will work on circuit 6 as shown in the drawing. This circuit will cover bedrooms, bath, and living-room lighting as shown. This circuit will require the use of the fixture-type mounting box as shown on materials drawing sheet (see page 100).

In the closet and hallway, this octagon-shaped box will be mounted on a ceiling joist. The bathroom light and the front-door lights will use this same type box mounted in the wall on a stud. Switches require the same type box as the wall plugs. Many people set the switch boxes 44″ from the floor to the bottom of the box, and the wall plugs 12″ above the floor.

Follow the number-6 circuit through the rooms; note that the switches in the bedrooms and living room are connected to wall plugs rather than ceiling lights. With this setup, lamps are controlled by these switches. If you desire ceiling lights in place of lamps, install an octagon box in the ceiling at the center of the room.

With all the boxes in place—fixture, wall-plug, and switch—you are ready to locate and drill all the holes for pulling the wire to all the boxes. You will still be using the same-size wire: 12-2 with ground, the same as for all 110-volt circuits. Some areas allow the use of 14-2, with a ground for lighting circuits. The 14-2 wire is a smaller wire and cheaper by a few pennies per foot. However, to get a good price on this size wire, you must buy a full box (250 feet) and then maybe use only half of it. Your best buy is to go ahead with 12-2. This circuit should also have a 20-amp breaker in the main panel. *Review the Light Fixture and Switch Layout floor-plan drawing.*

Now pull the wire through the holes drilled in the studs, double top plates, ceiling joists—whatever. When drilling holes, always look for the most direct route to the next box. Save drilling a hole anytime you can. Also, remember that by drilling a ¾″ hole you can pull two or three wires through this same hole.

First, pull your wire as shown on the floor plan, from front-door light switch to bathroom light switch. Leave the regular

6″ tail of wire out of both boxes. Pull the wire from the bathroom switch to the hall switch box and cut it off, leaving the 6″ tail.

Pull the wire from the hall switch to the front-bedroom switch box and leave a 6″ tail sticking out of this box. Pull the wire from the front-bedroom switch to the rear-bedroom switch through the box and out with a 6″ tail. Pull the wire from the bedroom box to the closet switch box, through the box and out with a 6″ tail. Your final run will be from the closet switch box to the main panel.

Next, snap in your 20-amp breaker, install your wire lock connector in the wall of the panel box and pull a 36″ tail into the panel. Split the outside covering on the 36″ tail to expose the black, white, and ground wire. Strip ½″ insulation off the end of both wires.

Next, place the black wire in the end of the breaker and tighten the screw firmly. Run the white and bare ground wires to the bottom of the panel box and connect under the screws as shown on the main-panel-box drawing (see page 102).

You will note that I constantly repeat: "Leave the 6″ tail, strip the wire, tighten the screw firmly." After hearing this over and over, you will soon perform these steps without thinking. Remember to keep all wires neat and orderly in the main panel box. You will see why after installing a few circuits.

Just outside the panel box, label the circuit Bedroom, Hall, Bath, Lights. Going out to the first switch box from the panel—the closet light switch—you'll find two 6″ tails. Strip these to the back of the box. Now run a wire from this box up to the ceiling box in the closet, leaving a 6″ tail on both ends, or boxes.

You now have three black, three white, and three ground wires in your closet switch box. At this time, refer to detail "D" on the switch-wiring drawings. Strip and twist all white wires together and install a wire nut. Then twist all ground wires together and install a wire nut. You have three black wires left: numbers 1, 2, and "S." Insert 1 and 2 in back of the switch as shown and connect the "S" wire under the screw. The "S" wire is your hot line to the next wall switch. Go through and wrap a piece of tape around your hot line coming

from the breaker. Put the tape around the hot line just before it goes into the switch box. Follow this wire "S" to the next box and tape—and so on to each box in the circuit. This will save a lot of confusion later along the circuit. All switches will wire up as above, except the last switch on this circuit, at the front door.

At this front-door switch, use a double box to install two switches: one for a wall plug and one for an outside light. (Refer to the Light Fixture and Switch Layout.) First wire up the switch and wall plug for the living room.

Review detail "C" in the drawings for this wiring. You will have two each of black, white, and ground wires. Make up as shown and connect the switch and outside light. Pull your wire from the switch box to the light-fixture wall box with a 6″ tail on both ends and strip both ends. At the switch, twist your one copper ground wire in with the two you have for the living-room switch. Connect the white wire with two on the living-room switch.

Next, insert your one black wire into the back of the outside switch at hole 2 as shown on the switch drawing. Cut a piece of black wire 4″ long, strip both ends back ½″, make a square bend on both ends to form a "U." Insert one end in hole 1 of the outside light switch; attach the other end to the "S" screw on the living-room switch. Your circuit is complete. Cover switches with duct tape. Remember to put a piece of identifying tape on your hot line before going into each wall box.

Garbage Disposal Wiring

Your disposal unit usually requires a 20-amp breaker, but check the instructions that come with the unit.

First, install a wall-plug box under the kitchen-sink location. Then place a box over the sink window for a light. Also install a double box in the wall 44″ up for the two switches, one switch for a sink light and one for a disposal unit. Pull a wire through drilled holes from the breaker in the main panel to this double switch box.

Next, pull a wire (12-2) from the switch to the plug box

under the sink. Connect your disposal switch as in detail "C" of the drawing on page 104.

Next, with your wall plug for under the kitchen sink in hand, break off the metal tip between the silver screws and the brass screws. This separates the two plugs. Insert your black and white wires 1 and 3 as shown in the drawings. Attach the ground wire under the screw at the bottom of the plug. Going back to the double switch box (for the disposal and for the light above the sink), drill a hole and pull a wire from the sink-light box down to the double switch box. Leave the 6″ tail on both ends. Take your one white wire and tie it with the two white wires from the disposal switch. Take the copper ground wire and connect it to the other ground wires.

Next, insert your one black wire coming from the sink light into a switch for the sink light. Use hole 2 as shown. Using another 4″-long black wire, formed in a "U" shape, insert one end into hole 1 of the light switch and tie the other end under the brass screw on the disposal switch. *Verify all conditions on job and with codes.*

Dishwasher Wiring

Again, review the drawings. A dishwasher usually requires a 20-amp breaker. Verify with the unit's instructions. Use the standard 12-2 Romex for this unit. When wiring the disposal, you used only one part of the wall plug under the sink. Now you will use the other half of this plug. Drill the holes or use the same holes used for wiring the disposal, and pull a wire from the main panel down to the wall plug under the sink. Insert the black and white wires into the bottom half of the disposal plug. Attach the ground wire to the screw at the bottom of the plug. This allows a separate circuit for the disposal and dishwasher. Staple all the wires to the stud within 12″ of every box.

Exhaust Fan and Hood

Over your cooking range, you may want to install a hood and exhaust fan. In some areas you must have a hood to pass the code. The standard range is 30″ wide, which means that you should buy a 30″ long hood. When selecting this unit, check the diameter of the vent pipe going from the hood up through the attic and roof. The metal pipe is usually 7″ in diameter and will require a roof-jack-type weather cap, mounted on your plywood roof before roofing. Install the cap before the roofing is applied over the plywood.

For the hood, install a wall-plug box above the hood location in a spot that will be hidden inside the overhead cabinet. Try 24″ down from the ceiling. Reviewing your wall-plug layout drawing, you will find this plug located above the range. As you will see, this plug is a part of circuit 2 and will be wired as shown on detail "A" but with only one set of wires.

Bathroom Exhaust Fan

If you have a bathroom that does not have a window, it will be necessary to install a ceiling exhaust fan. This fan must be vented to the outside through the roof, in the same manner as the kitchen hood. You should use a double switch box in this bath, one switch for light, one for fan.

Electric Water Heater

The average family should have a 40-gallon water heater to take care of their needs. When shopping for this unit, ask what size or amps are required for the circuit breaker to carry this size water heater. Your wiring supply house can also give you this breaker size. The standards are a 30-amp for the breaker and #8 3-wire cable. Ask questions. Also, your water heater will require 220-volt service.

Refer to the electric-service main panel drawing for the connecting of 220-volt circuits.

Electric Cooking Range

If you plan to use an electric range, this unit will require a 220-volt circuit, using a 50-amp breaker and #6 range cable, 3-wire and ground. This range cable is very costly. Measure the length you need carefully, then add 5′ to be sure. In some homes you can go through the crawl space, in others through the attic, but in some you must drill through the studs and pull the wire to the range.

When buying your range, also purchase a range cord to connect to the range and plug into the wall outlet. Purchase the exposed-type wall plug for connecting onto your large #6 range cable. Mount this plug on the wall about 12″ up from the floor. There are two different ways to go on this plug; see them at your supply house and decide.

Review your service panel drawing for wiring of 220-volt breaker.

Electric Dryer

This 220-volt appliance normally requires a 30-amp breaker and #8 3-wire and ground cable—the same as the electric water heater. Check this against the brand of unit you purchase or call your building department for the code in your area. (Remember, the building department is there to answer many of your questions; use them.) Measure very carefully how many feet of #8 cable you will need for the water heater and dryer, then add 6′ just in case.

Heating and/or Air Conditioning

Since there are many types of fuel and many types of heating units and air-conditioning equipment available, it would be impossible to discuss the wiring at this time. Select the most suitable type for your area. *Refer to the section on heating and air conditioning.*

Electric Washer

This appliance normally requires a 20-amp breaker on a 110-volt circuit, using 12-2 with ground wire. Snap in the breaker, knock out a slug inside the panel and install a wire lock connector. Drill holes through the studs and top plate to the wall-plug box behind the washer. Remember, the front of the box must be set out ½″ in front of the face of the stud to allow for drywall thickness. Next, pull the wire through the holes from the main panel box to the wall-plug box. Allow a 36″

tail in the main panel and a 6″ tail through the plug box. Strip the 36″ tail and connect the circuit breaker. Then install the plug in the box behind the washer.

Label the circuit, placing the name on the wire just outside the main panel box; do the same with all the other circuits. Now staple all the wires to the stud within 12″ of all boxes. Do not crush the wire with a staple. You can short it out if the metal staple cuts through the insulation on the black and white wires. Use staples *carefully* at every wire.

Main Service Ground

You now have many of your circuits installed and labeled for future reference. (We will come back to them later.)

Going back outside to your meter base, the main circuit-breaker panel, you must install your ground rod and ground wire. First, drive your 6′- or 8′-long copper-clad ground rod into the earth directly below the meter. Leave about 6″ of this rod above ground for installing the ground-wire clamp as shown on the Electric Entrance Service Drawing on page 101. Tighten this clamp to the ground rod securely. If your meter-base panel is installed between the studs, your ground wire will come down from the main box hidden inside the wall. At the floor line, this wire will be exposed 6″ out from the house to the rod clamp. For this type of installation, your solid copper ground wire will be encased in a metal protective cover. Verify this cover and wire size with your local supplier to make sure you are fulfilling the requirements of the code.

If you have installed a surface-mounted panel box, as shown in the drawing, your ground wire will be exposed outside the wall and must be protected by a conduit from the bottom of the panel to near the ground rod as shown in the detail of the drawing. Ask to look at one of your neighbor's installations.

Inside your panel box, you will note a built-in chart to list your circuits. Referring to the labels you placed on each circuit wire, list every circuit in proper order on this chart. In the future, at a glance, you will be able to flip any circuit desired instantly. Most codes insist that this chart be filled in correctly.

STANDARD ELECTRICAL MATERIALS

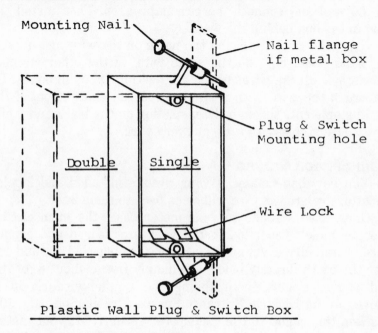

Mounting Nail

Nail flange
if metal box

Plug & Switch
Mounting hole

Double   Single

Wire Lock

Plastic Wall Plug & Switch Box

*NOTE: Nail these boxes to side of stud
or to side of ceiling joist.*

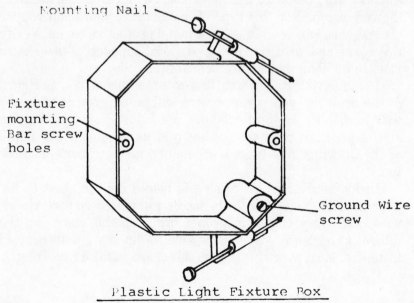

Mounting Nail

Fixture
mounting
Bar screw
holes

Ground Wire
screw

Plastic Light Fixture Box

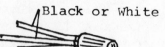

PLASTIC WIRE NUT

Black or White

2/3/4 Wire sizes

STAPLE

12" from Box

FIXTURE BAR

GROUND CLIP

GROUND CRIMP RING

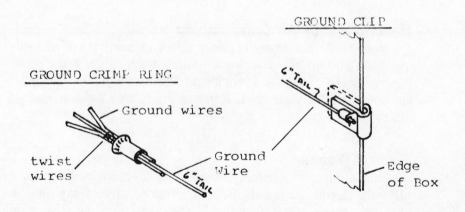

Ground wires

twist
wires

Ground
Wire

6" Tail

Edge
of Box

Most areas allow use of plastic boxes.
Verify all materials with local codes.
Plastic Boxes are easier to work with.

100

ELECTRIC ENTRANCE SERVICE

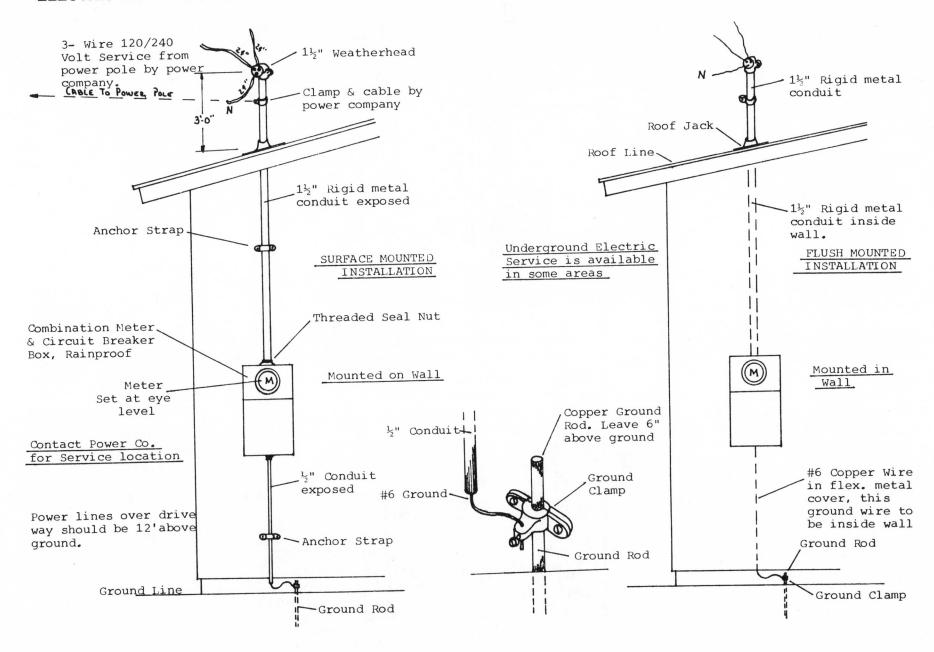

3- Wire 120/240 Volt Service from power pole by power company.
CABLE TO POWER POLE

1½" Weatherhead

Clamp & cable by power company

3'-0"

N

1½" Rigid metal conduit exposed

Anchor Strap

SURFACE MOUNTED INSTALLATION

Threaded Seal Nut

Combination Meter & Circuit Breaker Box, Rainproof

Meter Set at eye level

Mounted on Wall

Contact Power Co. for Service location

½" Conduit exposed

Power lines over drive way should be 12' above ground.

Anchor Strap

Ground Line

Ground Rod

½" Conduit

#6 Ground

Copper Ground Rod. Leave 6" above ground

Ground Clamp

Ground Rod

Underground Electric Service is available in some areas

N

1½" Rigid metal conduit

Roof Jack

Roof Line

1½" Rigid metal conduit inside wall.

FLUSH MOUNTED INSTALLATION

Mounted in Wall

#6 Copper Wire in flex. metal cover, this ground wire to be inside wall

Ground Rod

Ground Clamp

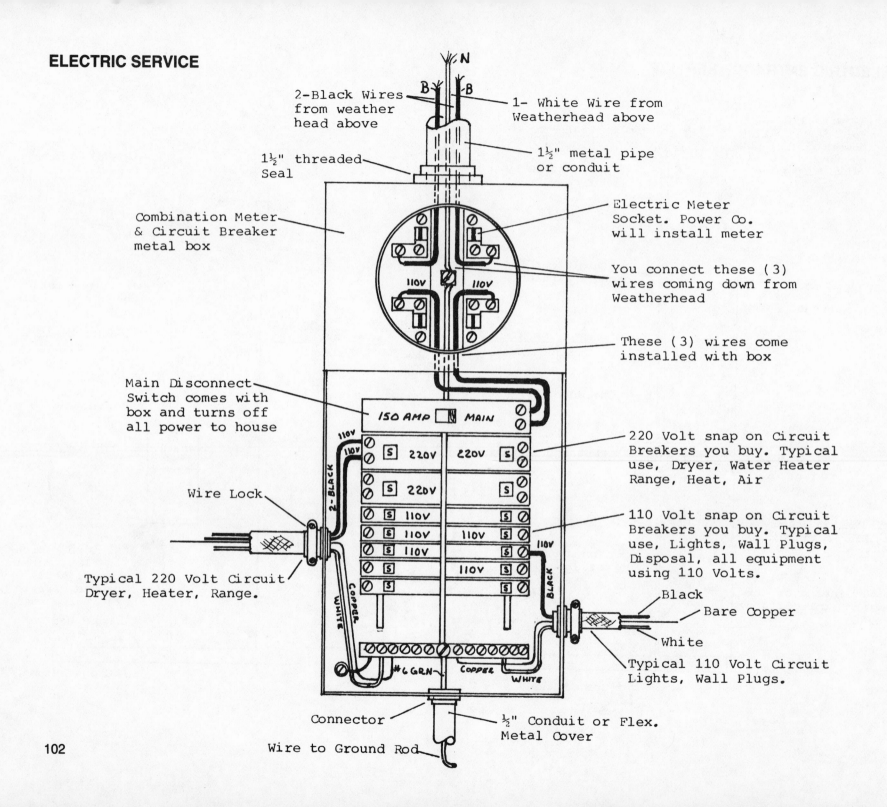

2-Black Wires from weather head above

B   N   B

1- White Wire from Weatherhead above

1½" threaded Seal

1½" metal pipe or conduit

Combination Meter & Circuit Breaker metal box

Electric Meter Socket. Power Co. will install meter

110V   110V

You connect these (3) wires coming down from Weatherhead

These (3) wires come installed with box

Main Disconnect Switch comes with box and turns off all power to house

150 AMP   MAIN

110V
110V

2-BLACK

S 220V   220V S

S 220V   S

S 110V   S
S 110V   110V S
S 110V   S
                   110V
S   110V S
S   S

220 Volt snap on Circuit Breakers you buy. Typical use, Dryer, Water Heater Range, Heat, Air

110 Volt snap on Circuit Breakers you buy. Typical use, Lights, Wall Plugs, Disposal, all equipment using 110 Volts.

Wire Lock

WHITE   COPPER

BLACK

Typical 220 Volt Circuit Dryer, Heater, Range.

Black

Bare Copper

White

#6 GRN   COPPER   WHITE

Typical 110 Volt Circuit Lights, Wall Plugs.

Connector

½" Conduit or Flex. Metal Cover

Wire to Ground Rod

WALL PLUG LAYOUT

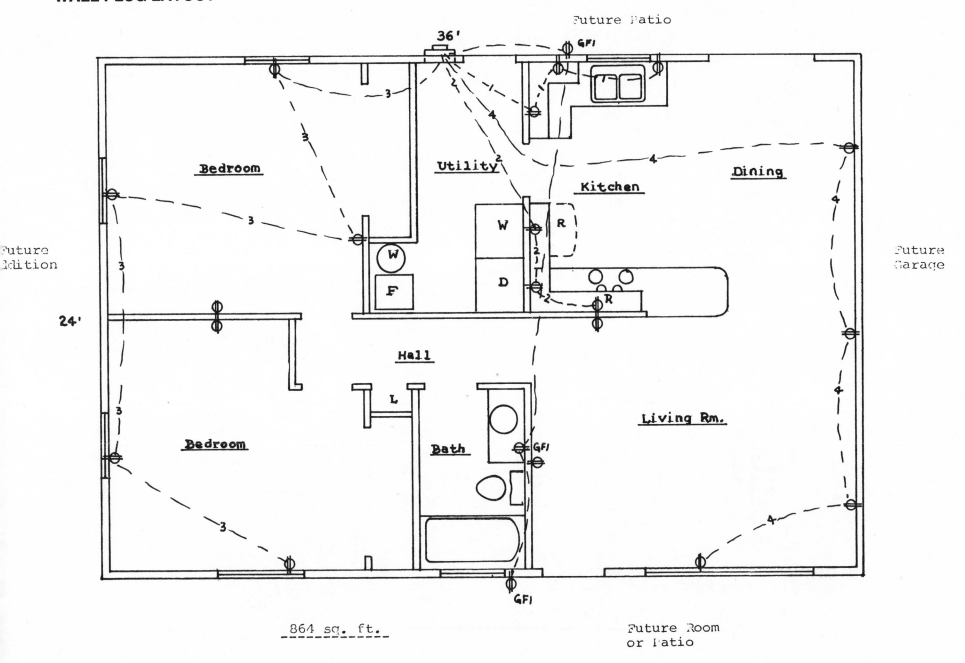

864 sq. ft.

Future Room
or Patio

WALL PLUG AND SWITCH WIRING

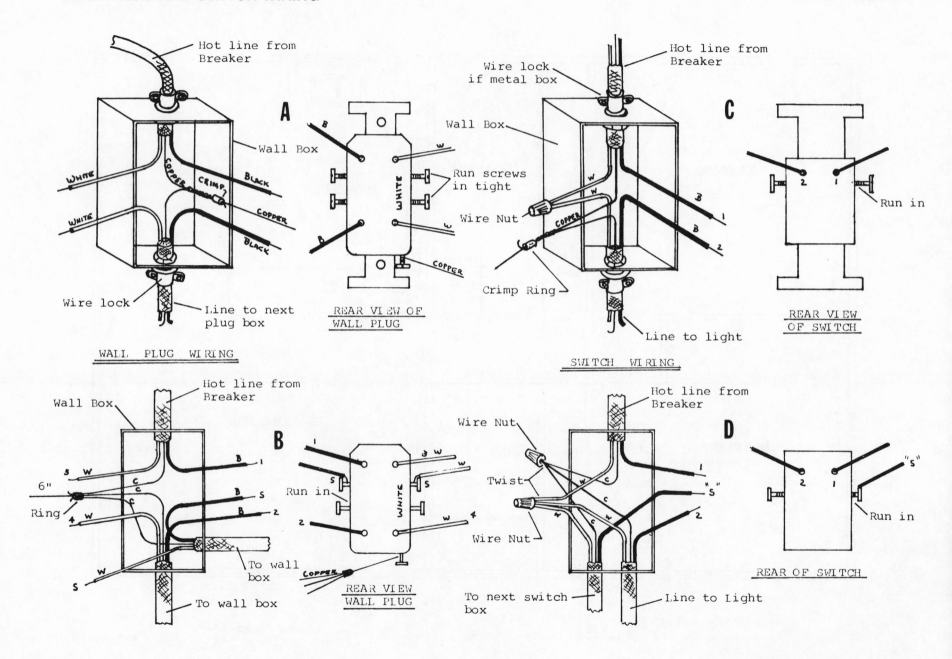

REAR VIEW OF WALL PLUG

WALL PLUG WIRING

REAR VIEW WALL PLUG

SWITCH WIRING

REAR VIEW OF SWITCH

REAR OF SWITCH

LIGHT FIXTURE AND SWITCH LAYOUT

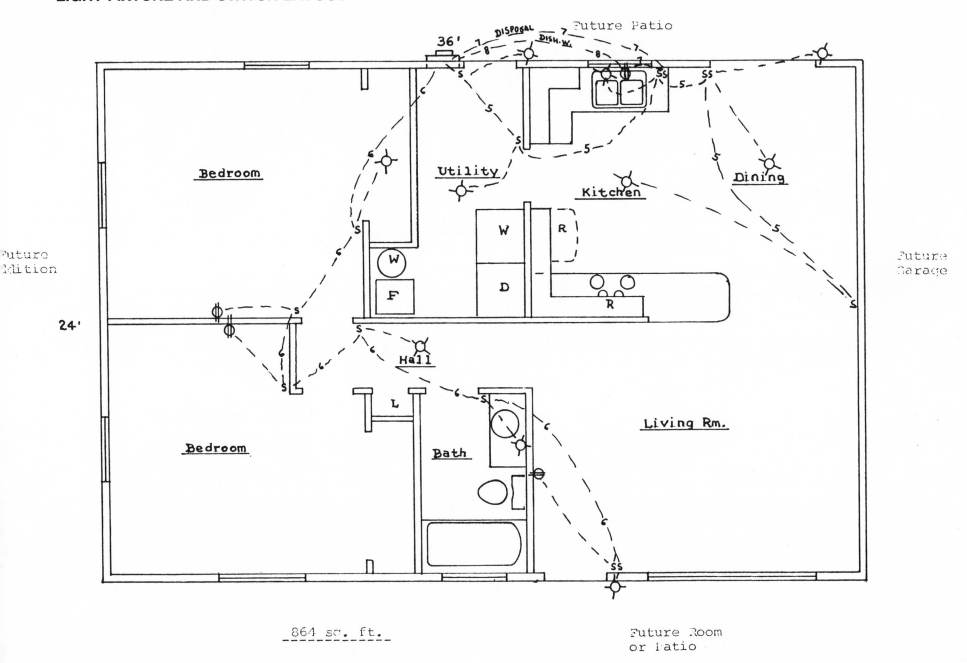

220-VOLT ELECTRIC LAYOUT IF ALL-ELECTRIC HOME

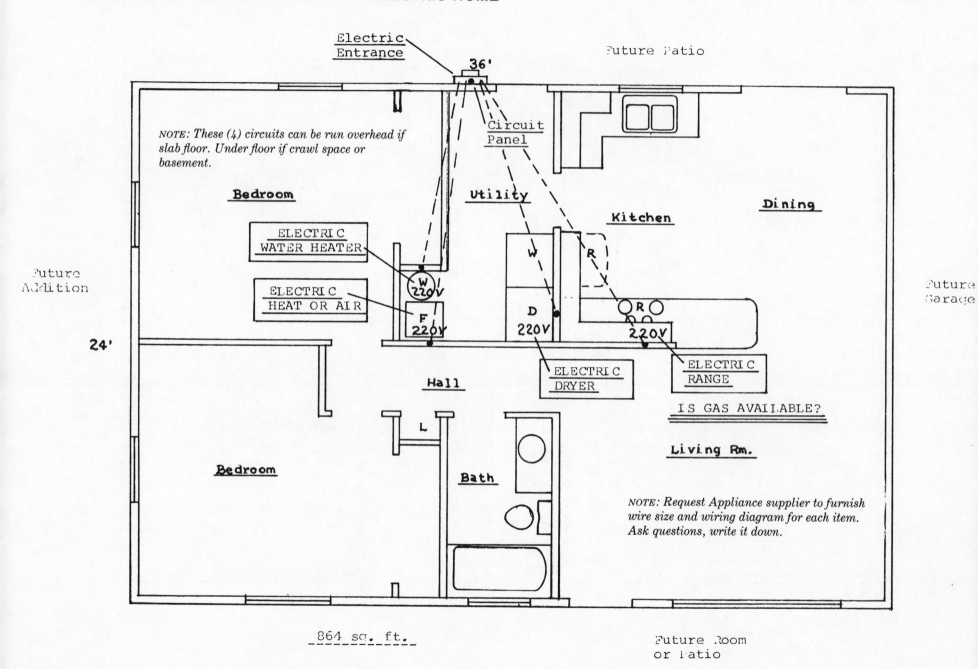

Electric
Entrance

36'

Circuit
Panel

Future Patio

NOTE: These (4) circuits can be run overhead if slab floor. Under floor if crawl space or basement.

Bedroom

Utility

Dining

Kitchen

ELECTRIC
WATER HEATER

W
220V

W

R

Future
Addition

ELECTRIC
HEAT OR AIR

F
220V

D
220V

R

220V

Future
Garage

24'

Hall

ELECTRIC
DRYER

ELECTRIC
RANGE

IS GAS AVAILABLE?

L

Living Rm.

Bedroom

Bath

NOTE: Request Appliance supplier to furnish wire size and wiring diagram for each item. Ask questions, write it down.

864 sq. ft.

Future Room
or Patio

HEATING AND AIR CONDITIONING

This will be a very difficult chapter, because there are so many types of fuel available. There are also many types of heating and air conditioning equipment for each of the many fuels.

It will be necessary for you to research all the available fuels in your area and select the most commonly used, economical heating. Perhaps you will not require air conditioning in your part of the country. This you must decide. When you have selected the most suitable fuel for your area, then you can shop for the heating equipment you will install in your home.

When purchasing this heating plant, ask your supplier how much of this system you can install yourself. Find out if he will give you a heating layout designed for your home, showing location and sizes of duct work and registers. Will he make a final check of your work and put the system in operation for you? Most suppliers will work with you in this manner and help you conserve.

Gas-Fired Forced-Air Furnace

Let us say that you have decided on a gas-fired, forced-air furnace to be installed in your utility room. First, set the furnace in place. With some units it will be necessary to build a platform about 24″ high to set the furnace on. From under this platform your furnace will draw air up into the furnace and around the heating element; then the fan will push the heated air out through the duct pipes to the various rooms.

The return air will probably be funneled through a large grille near the floor in the hallway. This will pull the air down the hall from all the rooms and the same air will be reheated over and over. With this type of furnace, you will need a 110-volt circuit from the main panel to the furnace to operate the blower and thermostat. Your supplier should provide the proper sizes and amounts of duct pipe, register boxes, and registers to complete the installation.

Electric-Fired Forced-Air Furnace

This type of furnace would be installed in much the same manner as the gas-fired unit. Since this unit would heat the air by electricity, you will have an electric heating element in place of a gas burner. You would then need a 220-volt circuit from the main panel to a fused disconnect switch on the furnace and then to the heating element inside the furnace.

Your furnace supplier will advise you what size circuit breaker to install and what size wire to use to carry this furnace. It would require the same size duct work, registers, etc.

Many brands of the above-mentioned heating plants are designed so that an air conditioning element can be installed right in the same furnace cabinet. In this manner, both heat and air will use the same duct work. If air conditioning is used, you will also need a separate 220-volt circuit to an outside cooling unit on the ground. You will also need a fused disconnect switch on this unit.

Roof-Mounted Furnace

You will find many types of roof-mounted units using gas or electric energy. These units can be installed with or without air conditioning and are used many times when space is not available inside the home. These units are self-contained and have all the duct work in the attic, just like the floor-type furnace. Again, if gas-fired, the unit will require a 110-volt circuit. If electric fired, a 220-volt circuit will be required, with a fused disconnect switch.

The supplier will advise you about the circuit breaker and wire size to carry this furnace.

Electric Ceiling-Cable Heat

Check with your local company to see if this is economical in your area. This method involves stapling a continuous, properly sized cable back and forth across the ceiling in each room. This cable is controlled by a low-voltage thermostat in each room so that each room can have a separate heating system. The heating cable installed on the ceiling is sometimes plastered in to hide and protect it. In some areas it is installed between two layers of drywall. Some types require each room to be completely wrapped in an aluminum-foil vapor barrier.

INTERIOR WALLS

WALL AND CEILING INSULATION

Wall Insulation

For insulating all the outside walls of a home, most people use the batt-type insulation. Batts without a paper wrap are called press-in-type batts. They are also made with a kraft-paper wrap or aluminum-foil wrap. The aluminum-covered in-sulation is slightly more expensive but worth the cost. Suppose you have selected the kraft-wrap, 4" thick, R-11 factor rolls of insulation for your home. To figure the amount of this insula-tion needed, measure around the outside walls of your home. Multiply this figure by the 8' ceiling height and this tells you how many square feet of insulation are needed. Your supplier will figure it from there. If you have an attached garage, insu-late the wall between the living area and the garage. Also in-sulate the outside walls of the garage. You will need a stapler, a sharp knife, and a tape measure to install your side walls. A stepladder will be needed for the ceiling insulation.

Along each edge of the batt, you will find a mounting flap to be stapled to the face of the 2×4 studs. Cut your batt the cor-rect length to fit tight from the top plate down to the fire block; press in place and smooth the batt out; then staple. You must have a neat and smooth job when finished. When the studs are less than 16" on center, and many will be, split the batt with a knife the long way to the proper width to fit tight between the studs. Studs should not be seen after the wall is all insulated.

Ceiling Insulation

Using R-19 batts, 6" thick, cut each strip the full length of the room; this is the same direction as the ceiling joists run. Start at one end, staple the batt in place, smooth and tight; work backward across the room. Do not insulate attic access door area. Fill every crack around windows and electrical boxes. All wiring, plumbing, and heating must be complete be-fore insulation is installed. Check insulation codes with your building department.

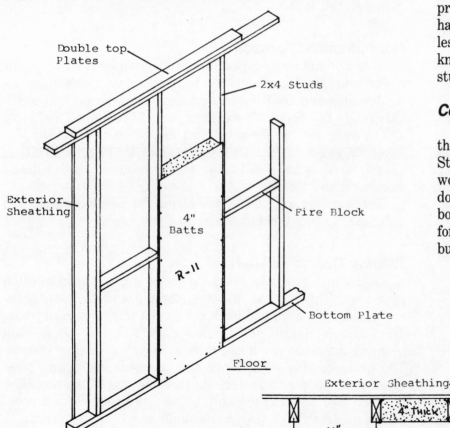

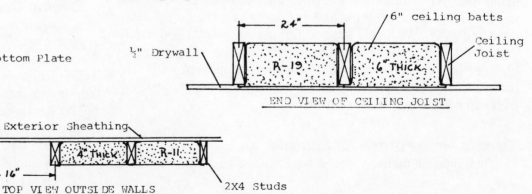

Wallboard Installation

Most parts of the country now use ½″ drywall board for covering all inside walls. This board is always 48″ in width and 8′ or 12′ long. If you have plenty of help, use the 12′-long sheets (they are heavy). You will need a sharp utility knife to score and break this board. You will also need tape, hammer, straightedge, and stepladders. Purchase a 50-lb. box of regular drywall nails; they have a cupped head.

Always install the lid, or ceiling, first; the wall drywall will then fit up tight against the ceiling drywall to help carry the load. Drywall sheets on the ceiling must be installed at right angles to the ceiling joists; never install a sheet running in the same direction as the ceiling joists. Measure from the wall out across the ceiling joists to the joist nearest to the end of the sheet. Mark this sheet at the center of that joist. Cut the sheet here with a knife and snap the sheet at the cut. Watch for ceiling-light boxes, vent fan or other openings you must make in the ceiling; cut them out now. You are ready to install the first sheet on the ceiling. With the sheet tight in the corner, nail it quickly. Drive a nail every 6″ across the sheet into every ceiling joist covered by this sheet of wallboard. Again, do you have all the backing in place? You must have a 2×6 nailed on top of every wall that runs parallel with the ceiling joists. Let this 2×6 lap over both sides of the wall into both rooms so you will have something to nail the end of the ceiling drywall to. These 2×6s for backing should be nailed in place before the plywood goes on the roof. Keep all the joints of ceiling drywall tight together.

When installing drywall on walls, always install the top sheet first (up against the ceiling drywall as shown on the drawing). Be sure that the end of every sheet is cut to fit to the center of the thickness of the stud. Watch for those wall-plug and switch boxes; do not cover them up. Nail wallboard securely, and always stagger the end joints as shown. When driving nails in wallboard, you must make a dimple or hammer mark at every nail to sink the nail and allow room for drywall mud to hide it. Do a neat job on your drywall; it will pay off later. To cut openings for boxes, make a firm knife cut to the box size, then cut an X in the opening; tap gently with your hammer; out it comes.

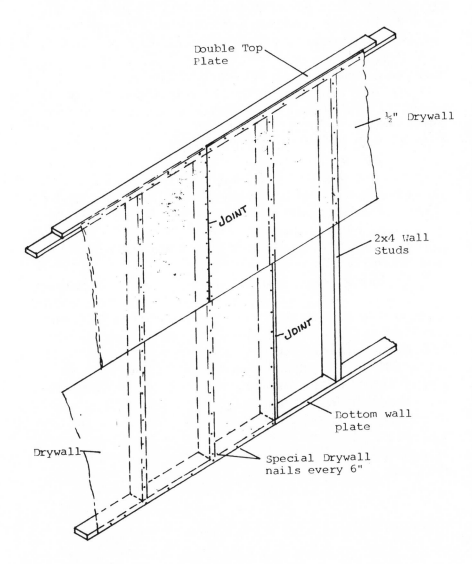

Double Top Plate

½″ Drywall

2x4 Wall Studs

JOINT

JOINT

Bottom wall plate

Drywall

Special Drywall nails every 6″

CORNER METAL INSTALLATION

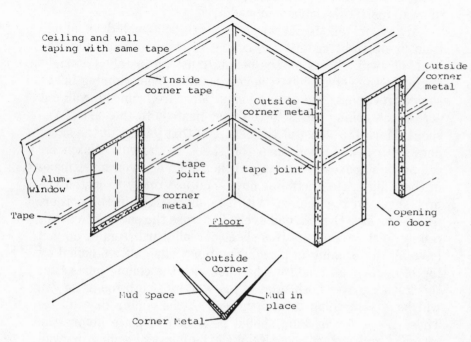

Ceiling and wall taping with same tape

Inside corner tape

Outside corner metal

Outside corner metal

Alum. window

tape joint

tape joint

Tape

corner metal

Floor

opening no door

outside Corner

Mud Space

Mud in place

Corner Metal

INSTALLING CORNER METAL

All outside corners require this metal to make a neat and clean edge. All inside corners require taping. If metal windows are used—without wood frames and inside trim—corner metal is used. All outside wall corners (shown above) use corner metal floor to ceiling. All openings without doors have corner metal on both sides of the wall thickness. Cut the metal with tin snips and nail it on the outside corners firmly with drywall nails. Do not hammer on the metal bead. In corners, such as those in a window, cut the corner metal to the exact length so that the bead of the metal touches in all corners.

DRYWALL JOINT TAPING

Now you have installed drywall throughout the house. You have installed the corner metal and are ready to do the taping of all joints.

Tools: You will need a 5"- and a 10"-wide putty knife to apply the mud. Also a plastic mud tray 4" × 12" for holding a supply of mud in hand. And then a sanding pad with a 48" handle and extra sandpaper. Purchase two rolls of special drywall tape. Drywall mud comes in a 50-lb. box; use the all-purpose mud; you will use several boxes throughout the house.

Start off by taping the joint, which is 48" above the floor, all around the room. Take a piece of tape long enough to run from the window to the corner. Place this tape in a container of water to soften it slightly. With a 5" knife put a band of mud along the full length of the seam; then apply the wet tape over the joint or seam and smooth gently with the 5" knife. Smooth out all excess mud now, or you will have a lot of work to sand it off later. Leave the tape showing now; tomorrow you will apply a 10"-wide band of mud over this seam. Keep it smooth at all times. Go over all nails two or three times; the first coat of mud will shrink when dry.

When taping an inside corner, such as those around the ceiling, form your wet tape into an angle to fit onto both walls; apply mud to both walls in the corner, smooth on tape, and spread out excess mud; let it dry. For a second coat, use a 10" knife for a band of mud on both walls of each corner. Apply the mud over the corner metal with a 10" knife; do not try to fill in perfect corners on the first coat; it will shrink and crack anyway. Take your time; it will take practice; you will soon learn to control each knife. You must keep all excess mud off the wall and keep it smooth. After the mud is completely dry, sand every joint, corner, and nail to reasonable smoothness for wall texturing.

EXTERIOR WALLS

BRICK OR STONE VENEER ON WOOD-FRAME WALL

Review wall-section drawings for proper foundation to support brick veneer load.

NOTE: Unless you have had some experience in laying brick, it would be wise to have this work performed by a professional. However, if you have the time to learn this trade, then do it yourself. Go to a nearby home where brick is being laid and spend some time. Watch every detail, note tools and materials used. These men learned the trade the same way— by watching, asking, and practice. You will need a mortar box and hoe, washed sand, and mortar cement. You will also need a trowel, a level, a chalk line, a hammer, and a brick-cutting chisel. If you have the time to do it, observe first.

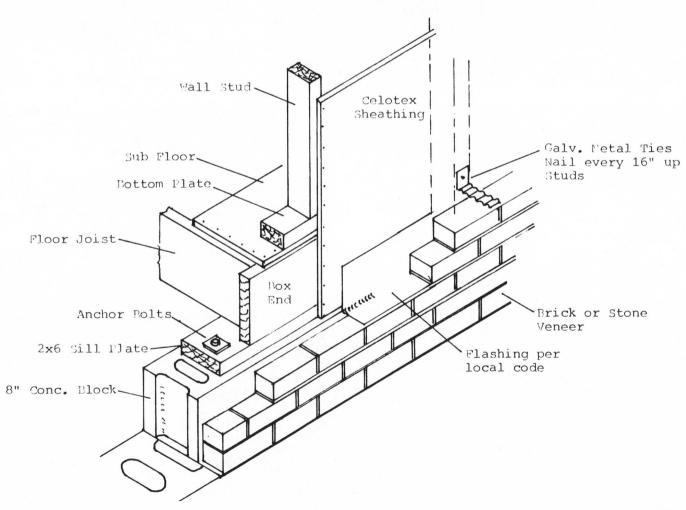

EXTERIOR SIDING

There are many types of good siding on the market today; it will be a matter of selecting the type you like best. There is the old-faithful lap siding in wood or a metal with baked-on finish. You will also find a variety of designs in the 4′ × 8′ sheets that are becoming very popular. This type of siding goes on fast, is also weather-resistant and fireproof, and takes paint well with little upkeep. Shop around before you buy. In some areas, Celotex sheathing is not required under this type of siding, but it is still wise to use it. The Celotex will help keep the heat out in a warm climate, and the cold out in cold areas. Most of these sheets are installed with #6 galvanized nails, every 8″ up each stud.

NOTE: If mostly 4′ × 8′ sheets of siding are used, a 1 × 4 trim board is used around all doors and windows.

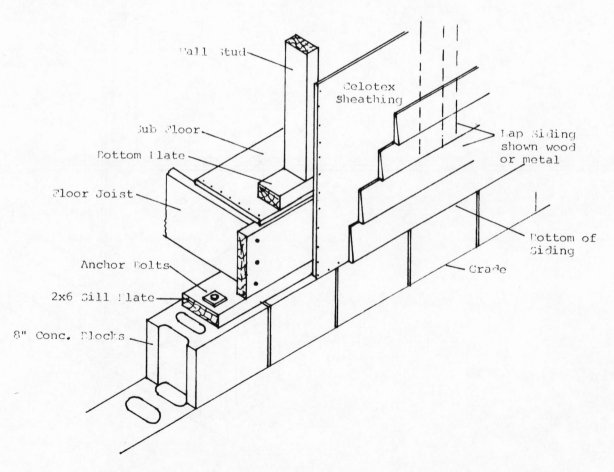

STUCCO-CEMENT PLASTER EXTERIOR

If your new home is to have a stucco exterior, the following guidelines should be followed. In most parts of the country, you will omit the sheathing or Celotex over the studs. Before you insulate your home, it must be wrapped with a tan weatherproof paper which is fully covered with chicken wire for bonding to the plaster cement. This wrap comes in 36″ widths, 50 feet in a roll. This wrap must be stretched tightly over the studs and nailed to them. It also requires a special nail every 6″ up every stud to hold the wire out from the paper to allow the cement to encase this wire. All outside walls should be wrapped as soon as possible after the roof is on. Only after wrapping the house can you insulate. Also: All drywalling must be completed before stucco is applied over the wrap. The constant pounding on the inside would destroy new green or uncured stucco. This type of exterior is usually applied in three coats: scratch, brown, and color-coat finish. This cement may be applied by gun equipment or by hand with a trowel. In most areas a metal weep mold is used at the bottom plate level (see detail below). After the home is completely wrapped, install the wire corners shown below. Around all doors and windows install the window metal or "J" metal shown below. This metal keeps the edges of the stucco neat and also serves as a guide to hold the thickness of stucco.

NOTE: Secure two bids from stucco contractors to do the plastering. This is a very critical step in completing your home. Make a deal with the contractor for you to wrap the house as he will require it. If you cannot save enough to make it worthwhile, let him do the complete job.

Wall Studs

Stucco Plaster

Weatherproof Paper w/chicken wire attached

Special Nails

Bottom of Stucco

Grade Line

Concrete Floor

Bottom Plate

Anchor Bolts

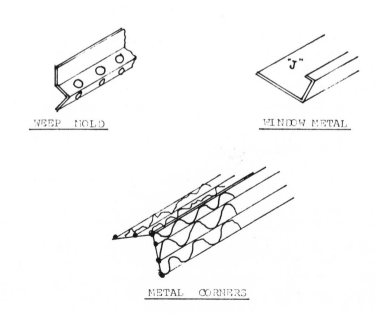

WEEP MOLD

"J" WINDOW METAL

METAL CORNERS

INTERIOR DETAILS

ACOUSTICAL SPRAYED CEILINGS

The drywalling is complete throughout the house. Every spot is sanded smooth. There is tape over every wall plug and switch. You are now ready to put the finishing touch to walls and ceilings. In your phone book you will find a contractor to do this work. Get a bid from two or more of them. The contractor will use a factory mix of acoustical material to shoot the inside of your home. He will use a large mixer and compressor to spray this material on your ceilings. Remember, you will not want this type of ceiling in the kitchen and bathrooms. These rooms will have an orange-peel texture like the walls and will take an enamel finish. This will make it easy to wash the ceilings and walls.

When the contractor starts on your home, be sure that he applies enough material on your ceiling to make it look just right to you. Go heavy but not too heavy. If too much material is used, it will crack or spiderweb when dry. When an acoustical ceiling is sprayed on, it is finished. No painting is required.

TEXTURED WALLS

The same contractor shooting your ceilings will also texture your walls and your kitchen and bathroom ceilings. To texture your home, this contractor will use basically the same equipment, with a different material mix.

Again, you must select the amount of texture you desire on these walls. Listen to his experience. He will give you a good surface. After this textured surface is completely cured out, you will then select paint colors for all your rooms.

Remember: Kitchen and bathrooms take enamel walls and ceilings over a primer or sealer. All other rooms should take a flat water-base paint on the walls. If you select a good brand of paint, one coat will normally give you a finished job. Two coats give it more depth.

Use a paint roller and pan for a good painted surface. Painting should not be started until all doors, frames, trim, and base are installed. All wood-trim doors, etc., should be painted with a color to match the walls, using a semigloss enamel, unless you prefer varnish on all woodwork.

SETTING OF DOOR FRAMES

At this time, you have ceilings shot and walls textured, and you are ready to set the doors. You will need your hammer, level, finish nails, and a small bundle of redwood shim shingles.

Earlier I recommended that you purchase prehung doors and frames. These units cost just a little more but they are well worth it. These doors will save you many hours of work and ensure a neat and well-hung door.

Set the assembled door and frame unit into an opening of the proper width. Be sure you have this unit set in the opening properly so the door will swing into room as desired.

Next, move the door frame, hinge side, over tight against the wall stud. Make sure the edges of the door frame are flush with the face of the drywall in both rooms. With #8 finish nails, top, bottom, and middle, anchor the door frame, hinge side to the wall stud. Do not drive nails fully into the frame at this time.

You now have a crack along the opposite side of the door frame. If you kept to your previous door-frame dimensions, this crack will not be too wide. If necessary, install a strip of drywall full length to help fill it up. With the door in the closed position, tap in a shim shingle between the frame and stud to adjust the crack between the door and frame. Use a shim shingle up and down both sides of the door frame if it is needed, but do not bind your door to a tight fit.

Anchor both sides of the door frame with finish nails but

watch the door crack. You are ready to install the trim up both sides and across the top.

Do this on both sides of the door, in both rooms. Study the door trim where you now live, if you are not sure how it is installed. Holding each trim board in place along the correct side of the door, mark it at the top for making a 45-degree cut. You will need a cheap wooden miter box for making these angle cuts around all the doors. Nail these trim boards to the edge of the door frame with #4 finish nails about every foot along the full length. Using a nail-set punch and hammer, sink all the nails slightly in the trim and frame. Using wood putty or spackle, fill these nail holes and sand them after they dry. This door and frame is ready for enamel or varnish finish.

KITCHEN CABINETS

There are a number of factors to be considered when purchasing kitchen cabinets.

Cabinets can be made on the job, built right in the kitchen, job-measured and built in a local cabinet shop, or purchased in sections from a large supply house as factory-built, assembled and unassembled, prefinished or unfinished. You may want to get some prices for each of the above types, but you will find the cheapest way to go will be the factory-built, unfinished cabinet sections, unassembled. Looking at suppliers' sample units you will find that most of these are well-built. Also, these cabinets are much less expensive than custom-made cabinets.

Review the cabinet detail drawings (pages 116–117) to acquaint yourself with the various details you must consider in designing your own cabinets.

No doubt the blueprints of your home indicate a complete layout of the cabinets, showing sink and range locations. From the sink location, you have already framed in your window in the proper place. With crayon, outline your base or lower cabinets on the floor. Draw a line out 24″ from the wall. This will be the front face of all the cabinets.

Next, locate the center of the sink window on the floor. Measure 18″ each way from this line to outline the sink, directly under the window. This area will allow for a 36″ base-cabinet sink-front unit.

Next, identify your range location. Will you purchase a standard free-standing slide-in range? If so, you need a 31″ space in the base cabinet area for the range to slide into. Draw the range on the floor. If you plan on using a countertop range, recessed into the top of a cabinet, you will need a 30″-wide base cabinet. Locate this cabinet on the floor.

Factory-built base cabinet units can be purchased in many widths to fit any space you have allowed for your cabinets: 12″, 16″, 18″, 24″, 30″, 36″, 48″ widths.

On the floor, with crayon, design your cabinets. Locate your drawer and shelf units. Your supplier will furnish you with a chart showing every dimension of all units they can supply, both base cabinets and overhead units.

Refer to cabinet details in this book. They will guide you around sink, range, hood, refrigerator, etc.

If your home will have a wood floor, it will be best to install your double or finish floor before installing cabinets.

You have now purchased your cabinets and are ready to install them.

First, set all base cabinets in place according to the lines on the floor. With all the sections sitting tightly together and with the face or front of the cabinets aligned, drill a hole through the rib of one unit. Insert a wood screw and tighten it to attach this unit to the next base unit. Screw all the base units together around the kitchen. Push the entire base unit against the wall and into the corner. When they are tight and in place, drill through the back rib of the cabinets in line with the wall studs; insert the screws and anchor the base to the wall. Make sure the joints on the face of the units are flush.

Overhead Cabinets

If the cabinets include a corner unit, set the corner overhead unit first. With the unit in place and up tight against the soffitt, drill holes through the inside back of the cabinet and into the studs. Insert a wood screw and tighten the unit to the wall. Install the adjoining units in each direction from the corner in the same manner.

Formica Counter Tops

There are many designs and colors of Formica to choose from for your cabinet counter tops. You should select a counter with a rolled edge on the front and a backsplash attached on the rear edge. Your best buy on Formica tops will be at a large store which carries various lengths and colors in stock. If your cabinets have a corner, you will need to purchase two pieces of top with one end of each cut on a 45-degree angle to fit into the corner. Be sure to get bolts to tighten this joint together.

However, before joining this angle together, you should make your cutout in the top for the double sink. This sink normally comes with a sink-mounting ring and full instructions for cutting the proper-size sink hole in the Formica top. Follow these instructions closely. You will need a saber saw to make this cutout. Install the sink and ring into the counter top before installing the top on the base cabinet. Anchor the top to the base cabinets from inside up with wood screws through the base ribs.

If your cabinets are unfinished, there are a number of finishes to choose from: enamel, varnish, stain, then varnish. Your local paint store will guide you.

KITCHEN CABINET DETAILS

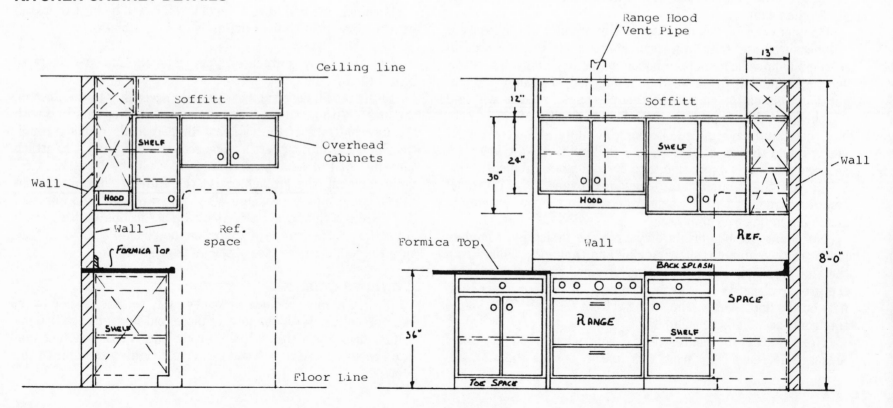

KITCHEN CABINET DETAILS (cont.)

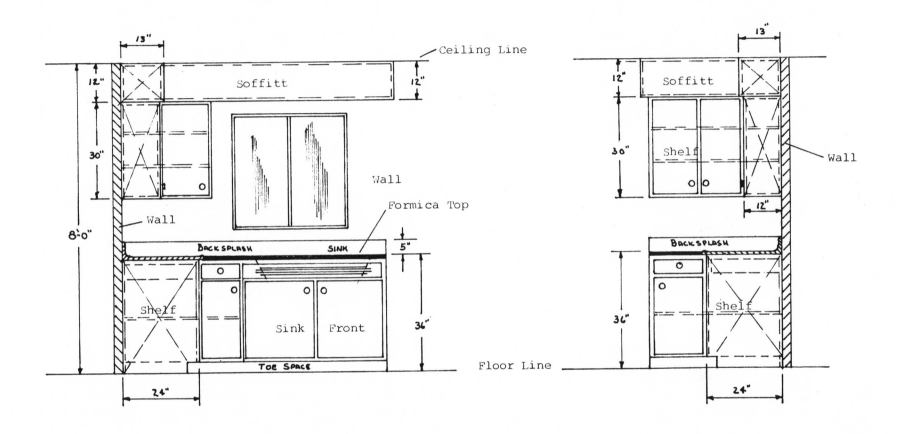

FLOOR PLAN OF CABINET LAYOUT

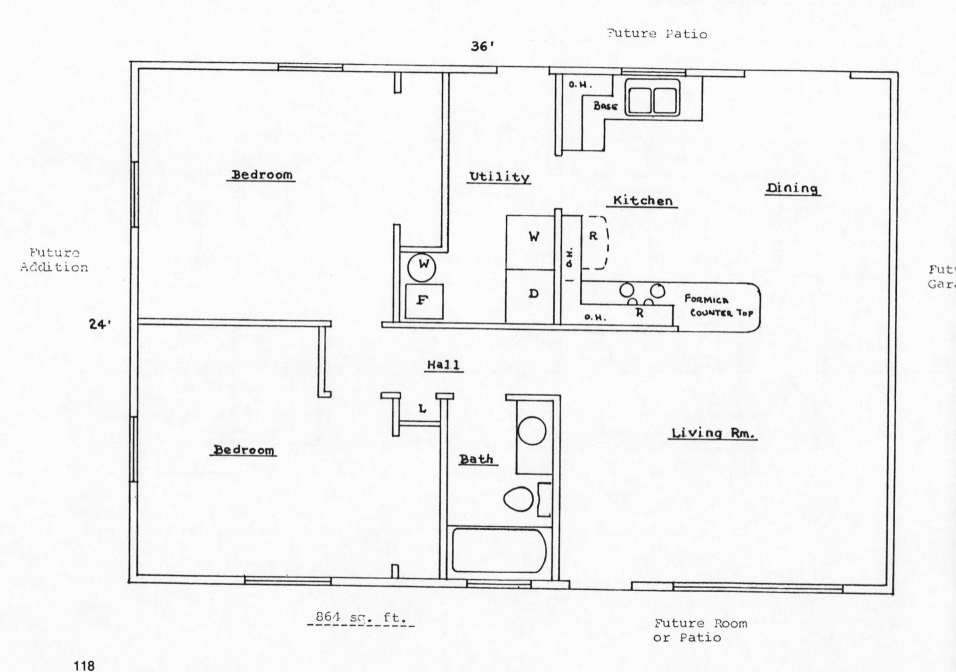

Future Patio

36'

O.H.

BASE

Bedroom

Utility

Dining

Kitchen

W   R

Future
Addition

W

O.H.

Fut
Gar

F

D

R

FORMICA
COUNTER TOP

O.H.

24'

Hall

L

Living Rm.

Bedroom

Bath

864 sq. ft.

Future Room
or Patio

OUTSIDE DETAILS

FINISH GRADING OF YARD

With the exterior of your home all completed, siding installed, trim painted, you are ready to put the finish grade of your yard around the home. If much earth is to be moved, you may need to hire a tractor with a blade on the rear to level out your yard. In some areas you can rent this tractor and do the work yourself. Check it out and go the cheaper way.

On this plot plan, you will notice the arrows around the house. These arrows indicate the direction water should flow during a heavy rain. In most areas, your finish grade or earth must be 6″ below the wall bottom plate or any wood members of the house. The swales shown on this drawing are proposed water channels to direct water to the street. You should hand-rake your yard to slope gently down to the swales. Then slope it down from the property line to the swale. (You are not allowed to drain your yard over onto the neighbor's yard.) You will note that the side yards and front drain toward the street. The rear yard should drain to the rear unless this land is higher. If so, drain the rear yard to a swale across the back of the house, then through the side yard and out to the street. This is the time to get your yard in perfect condition and be done with it.

LANDSCAPING

Shade trees are very important, not only for appearance but to shade your home. These trees should be planted with much thought given to the morning sun from the east, and the after-noon sun in the west. Your home will stay cooler in summer if your trees are set in the best location. When selecting shrubs, bushes, or plants, check out the size they will be when fully grown. Try for larger shrubs on corners with lower types under windows. If you will have a grass lawn, follow the instructions of your local seed supplier as to which grass is most commonly used in your area.

119

GLOSSARY

ACCESS WINDOW   Usually a standard basement window, steel or aluminum with three panes of glass. This access is required only with wood-floor homes for entrance into the crawl space under the floor.

ACOUSTICAL CEILINGS   A factory-mixed material which is mixed with water on the job and with high pressure equipment is sprayed on ceilings directly over taped and filled drywall. Serves as a finish ceiling and improves acoustics in the home. Do not use on kitchen and bathroom ceilings; these ceilings should be textured like all walls and have an enamel finish for washing down.

ANCHOR BOLTS   A ½"-diameter steel bolt, 10" long, with threads at one end and a square bend at the other. This bolt is set in freshly poured concrete, leaving 2" of the threaded end showing above the floor. These bolts anchor the outside walls of the home to the concrete footings. See footing details (page 47) for spacing of anchor bolts.

BACKFILL   Only required on homes with basements. In basement details (page 44), note that the hole dug in the earth is at least two feet wider on all sides of the basement than its final dimensions will be. This allows room to install the footing drain and for waterproofing all outside basement walls which will be below ground. Only after the home is completely framed can the earth be pushed in to backfill around the outside of the basement walls. If backfill is attempted before framing, it is possible to push in the upper portion of a basement block wall.

BATTERBOARDS   Refer to drawing (page 27). This is the very first step in laying out your home. These boards, with stakes and chalk line, are required to establish the outside dimensions and squareness of your home.

BEARING WALL   All outside walls are considered to be bearing walls. If roof trusses are used, the walls under both ends of the trusses are bearing walls. If regular ceiling joists are used, walls under both ends of the joists are load-bearing walls. These walls also carry the roof load, rafters, sheathing, and roofing. A concrete-floor home not using roof trusses must have a load-bearing wall through the center of the home to support the weight of the ceiling and roof. This bearing wall is supported by a concrete footing under the slab floor. If the home has a wood floor and crawl space or basement, use a built-up wood beam or steel I beam to support floor, ceiling, and roof loads. This beam or girder is supported by concrete piers for a crawl-space home, and adjustable jack posts in a basement home. Refer to drawings.

BLOCKING   Required between floor joists, ceiling joists, and roof rafters. If floor joists are installed 16" on centers, cut 14½"-long blocks to nail on edge between the joists. These blocks will be cut from the same size lumber as the floor joists. If ceiling joists and roof rafters are spaced on 24" centers, cut 22½"-long blocks as shown on drawings.

BOTTOM PLATES (Wall)   In a concrete-slab home, bottom wall plates will be same size as wall studs, 2×4 in most areas. Bottom plates under all outside walls of a home must be treated lumber to resist moisture, decay, and termites. Lumber yards carry treated plates. All inside-wall bottom plates require regular 2×4 lumber. Homes with crawl space or basement require regular 2×4 lumber bottom plates throughout because of the metal termite shield.

BOTTOM PLATES (Foundation)   On top of the block wall for a crawl-space or basement home, a 2×6 board is usually required. This plate need not be treated because the metal termite shield is installed directly under the bottom plate. This bottom plate is carefully drilled to fit down over anchor bolts installed in the top of the block wall.

BOX END PLATES   The same size lumber as the floor joists, this plate stands on edge and is nailed to the end of every floor joist, usually at the front and back walls of the home. Plates are then nailed securely to the bottom plate on top of the termite shield.

BUILDING CODES   Almost every city or county has a somewhat similar building code. These codes assure you that you will have a safe and sound, well-built home. These codes will serve to guide you in the building of your home. If a contractor is building for you, it will assure you that he has built your new home properly. The greatest variation in codes throughout the country will be in the foundation and footing area. This is due to the variations in climate in different parts of the country: frostline, depth of ground freeze, etc.

Obtain a copy of your local building codes from your city or county building department. Study these codes well, and remember, you are not just learning how to build your own home, you can be learning a completely new and profitable profession. If you were to thoroughly learn just the Glossary of this book, you would do well on a State Contractor's License test.

BUILDING DEPARTMENT This is an office of local government. In many cases, your building-permit fee supports the operation of this office. This office issues your permits and makes inspections at various times during the construction of your home. *Do not be afraid to ask them questions;* they are the experts. The purpose of their office is to see that your home is built properly. Let them guide you from the start.

CIRCUIT-BREAKER BOX OR METER BASE This electrical box is usually mounted on an outside wall as directed by your local power company. It may be in the wall or surface-mounted and weatherproof. In many areas, this box includes the meter base for the power company to install the meter. From the top of this box, going up above the roof to code height, you will furnish the pipe conduit, weatherhead, and entrance cable: two black wires and one white wire to which the power company connects electrical service. Inside this box, you will have a main disconnect switch to shut off all power into your home. Below this main switch, you will snap in circuit breakers of various amp sizes which supply power throughout the home.

CIRCUITS (Electrical) From the circuit-breaker box, you will run many circuits for lights, wall plugs, furnace, water heater, range, etc. (Refer to details in the electrical chapter.) Every type of circuit requires a circuit breaker of a certain amp size. Review the chapter for complete details.

CLEANOUT (Plumbing) In case of blockage in waste or drain lines, a plumbing fitting, "Y," "T," "L," or a coupling should be installed at the end of all waste or drain lines. These fittings should include a screw plug or cap for ease of cleanout under sink and bath lavato-

ries. If a septic-tank system is used, install a cleanout plug between home and tank. (See plumbing chapters for complete details.)

CONCRETE CAP BLOCKS This solid concrete block is 4″ thick, 8″ wide, and 16″ long. When required, this cap is installed in mortar on top of the block wall for crawl space and basement homes. Most areas do not require this cap. Verify this with your local block supplier or local codes. If your home will be financed through FHA minimum property standards, this cap will be required.

CONCRETE PIERS Along with the blocks, these are used to support the girder through the center of the home. These piers carry a large portion of the weight of floor, walls, ceiling, and roof structure. Piers are also used to support roofed patio posts.

CORNERS, WALL FRAMING All outside corners of all walls throughout the home must have this 1×6 board brace. All inside walls connecting to outside walls at a right angle must have this brace at the outside wall end. This 1×6 brace runs from the top plate down at a 45-degree angle through the bottom plate. This 1×6 brace is notched into the top and bottom plates and into every stud it crosses. It is anchored with three 8-penny nails at every stud and plate.

CRAWL SPACE Under all wood-floor homes there must be a crawl area, usually with a 24″ clearance. This allows for future repairs or changes in wiring, heating, or plumbing. With proper ventilation in this area, dampness can be kept from under the floor.

CRIPPLES (Framing) These boards are cut short to fit between wall studs above and below window openings and above doors from headers to top plate.

DRAIN "P" TRAP Required under all sinks, tubs, showers, etc. This fitting is designed to hold a full pipe of water in the lower portion of the fitting. This water serves as a seal to avoid the danger of sewer gases coming back up through tub or sink.

DRYWALL BACKING On top of all walls running parallel with the ceiling joists and the end of the house gable walls must have backing. On top of parallel walls, nail on a 2×6 flat with equal amount of 2×6 projecting over on both sides of wall. This allows proper nailing of drywall to ceiling. On gable end walls, install 2×4 on top of top wall plate for proper nailing of drywall ceiling at both ends of home. This backing must be nailed in place before installing roof sheathing plywood, due to lack of space for proper nailing.

DRYWALL MUD This is a factory-mixed compound used to hide nails and joints in drywall after installation. Every nail must be recessed into the drywall; leave a slight hammer-mark dimple at every nail. Using a 5″-wide putty knife, fill each dimple with mud. When the mud is dry, go over every dimple again; mud shrinks when drying. Any excess mud on wall or ceilings will take a lot of extra work to sand off later. Before tape is applied over all joints, apply a coat of mud over the joints; next, place tape in the mud over the joints and wipe the joints with a 10″-wide drywall knife to remove all air bubbles and smooth out all excess mud. It is all right to let the tape show through. After the joints are dry, apply another coat of mud with the 10″ knife. You must keep mud smooth.

EASEMENT This is a portion of your land that you own but cannot build on. In some cases it is needed for the installation of various utilities: underground gas, phone, and electric lines. However, many parcels do not have any type of easement. When purchasing your land, be sure that you find out about any easements that would affect the size and shape of your home. Check it out before you buy.

EXPANSION JOINTS An expansion joint is a divider between two concrete slabs and is made of a ¾″-thick Celotex sheathing cut in 4″ strips and placed on edge between slabs. This joint allows the expansion and contraction of slab and earth without cracking slabs. It is most often used across driveways, sidewalks, and other large areas of concrete. Most concrete-slab floors for homes do not require this joint. A basement-floor slab should be separated from the block walls by an expansion joint.

FHA An FHA loan means that the Federal Housing Administration, a branch of the federal government, is insuring your mortgage to the lending bank. FHA has nationwide Minimum Property Standards, or building code, which must be met. This book attempts to meet these codes in every way possible.

FINISH FLOOR LINE This is the top surface of the floor, ready for carpet or tile. This floor must be, in most areas, 6″ above the earth for concrete slab and 8″ above the earth or finish grade for wood floors. Check this with your local codes.

FINISH GRADE This is the final level of the earth or yard around your home. This is achieved by hand-raking out from the walls. The yard must have a slight slope down and away from the walls to guide surface water away from the home. (Study plot plan drawing.)

FIRE BLOCKS Used to brace walls as well as to block off possible fire channels between studs. These blocks are made from scrap 2×4 studs and should be installed halfway up the wall between all studs. If the studs are installed on 16″ centers, these blocks will be cut 14½″ long. Stagger these blocks as shown on drawings for ease of installation.

FOOTING DRAINS Installed around the outside of basement footings or foundation in wet areas to help prevent a damp basement. This drain is usually connected to a sump pit with pump. Check codes in your area if a basement is planned.

FOOTINGS (Foundation) Poured concrete base around the entire home. These footings carry or support the weight of the home. The width and thickness of these footings is determined by the type of construction of the home and the climate in your area. In a cold climate, the depth of these footings is determined by the frostline or the depth at which ground freezes in the winter. Your local codes show this depth.

FORMS, CONCRETE FOOTINGS Homes with crawl space or basement usually have trenches dug into the earth to the proper size of the required footings. This trench is then filled with concrete. Homes with concrete-slab floor require, usually, a 2×6 board on edge around the outside dimensions of home. Concrete is then poured to the top of this level form board. This method meets the code in most areas which requires 6″ from finish floor to finish grade. These form boards must be securely staked and anchored to the ground before pouring concrete.

FRAMING Term used when studs are nailed in place between the top and bottom plates to create a wall of the home. This term also covers standing the walls in place and nailing them together ready for the roof.

FROSTLINE The depth at which ground freezes in winter in a particular area. This determines how deep the footings can be set in the ground. The frostline also determines how deep water lines can be placed.

GIRDER   A wood beam built up of three 2×8's bolted together, or a 6″ I beam running the full length of the home, installed under a wood floor for a crawl-space or basement home. This girder carries the weight of the floor and walls, and a portion of the roof weight. For crawl-space homes, the girder is mounted on concrete piers and blocks. For basement homes, the girder is supported by adjustable jack posts.

GROUND PLUMBING   This term is usually associated with a concrete-slab floor. All drain or waste lines, along with hot and cold water lines, are installed in the ground below the concrete floor slab before the concrete floor is poured.

GROUTING   A thin, pourable mixture of cement and sand which is used to fill the cores in a block basement wall. This mixture, when poured into a block wall with steel rods, makes a strong, reinforced, one-piece wall. Grouting and re-bars are also used in retaining walls.

HEADERS   Heavy wood beams built into the walls over window and door openings. These headers help carry the roof load over openings. Example: A 4′-wide window would require a 4×6 header.

HOSE BIBB   An outside water faucet for connecting a lawn hose. In a cold climate, a special faucet is needed to avoid freezing in winter. The average home should have at least two outside faucets.

INSPECTIONS   These are made by your local building department, usually the same people that issued your building permit. Verify with them what inspections they make at various stages of construction. Do not hesitate to ask the inspector for an explanation of any detail they require. Let them guide you.

JACK POST   Required in basement homes only. This adjustable jack post is made of heavy steel pipe with a threaded screw in one end and a steel mounting flange on the bottom for anchoring to the basement floor. This jack is located every 8′ under the built-up girder or 6″ I beam. The jack post allows you to level the top of the girder on which the floor joists rest.

METER BASE (For Electrical Service)   Usually built in as a part of the circuit-breaker box. After the final inspection of your home, the power company will plug into the meter and turn on power to your home.

PATIO ROOF POST ANCHORS   These post anchors are usually H-shaped galvanized metal. Insert this anchor into the patio slab and piers at the time the patio is poured, leaving the top half of the H exposed so the bottom of the 4×4 post can be installed. See patio details.

PATIO SLAB   This slab is usually formed with 2×4's on edge to give a 4″ thickness. The top of the patio slab should be 2″ or more below the main living-area floor to avoid possible water run-in. Concrete piers, at the outside corners and every 8′ along the full length, should be poured at the same time as the slab for the 4×4 posts that will support a future patio roof. After the forms are ready, dig out the earth 16″ square and 12″ deep at each post location. Extra foundation is not required under the rest of the patio slab. If the patio roof will be installed in the near future, install metal 4×4 post anchors at this time (see below).

PENCIL ROD   This material is required in some areas and should be installed only in basement walls and retaining walls. Two of these ³⁄₁₆″ steel rods are laid horizontally on top of each third row of blocks. Then mortar and another row of blocks are laid. These rods are thus encased in the wall for added strength.

PLATE STRAPS   Galvanized steel strips which come in various lengths. Where sections of the top and bottom plates of the walls are butted together, you must install a plate strap on both sides of the plate to strengthen the joint. *A must.*

PLOT PLAN   A complete layout of the entire property, showing the location of home, drive, walks, trees, easements, and every dimension and legal description covering your property. This plan should also show finish-floor and finish-grade elevations.

PLUMBING WALLS   After all walls are standing in place and anchored to the floor, you must plumb and line each wall. With a 48″ level to check with, push the top of each wall in or out until it is vertical. With a brace from the floor to the top of the wall to hold it in place, the bottom of the wall is anchored to the floor in a straight line. The top must be held in a true straight line also. This is checked by attaching a chalk line along the top of the wall on the outside and pushing the wall in or out and bracing it on vertical.

RAFTERS   Usually 2×6 boards installed on 24″ centers at the desired roof pitch or slope. Plywood roof sheathing is then nailed on top of the rafters. If roof trusses are used in the home, the top chord of truss becomes the rafter and the bottom chord of truss is the ceiling joist.

REINFORCING BARS (Horizontal) These bars are most commonly used in the concrete footings or foundation of the home. In home construction, the standard used is ½″ diameter or #4 steel bar. This bar has a rough exterior to lock in and strengthen the concrete footings. Normally two #4 bars are embedded in the concrete footings if required by your local codes.

REINFORCING BARS (Vertical) This installation is only required in basement block walls and retaining walls. These ⅜″ diameter or #3 steel bars are embedded vertically in the concrete footings and extend the full height of the block wall. Standard spacing is usually every 32″. The cores of the blocks holding these bars are then filled with concrete grouting.

ROOF JACK Every plumbing fixture must be vented through the roof. Bath fixtures are usually all vented into one 3″ pipe through the roof. The kitchen sink has a separate 2″ vent. Your local lumber company carries roof jacks which slip down over the vent pipe and are nailed to the plywood roof before the roofing is installed.

ROOF TRUSSES If roof trusses are used, they take the place of the ceiling joists and roof rafters. You will find that trusses go up much more quickly and are less expensive than joists, rafters, and bearing-wall foundations. Check out both methods.

ROUGH-IN HEATING After your home is completely framed, with all walls in place and plywood sheathing on the roof, install the furnace or heating plant. Then install the ducts or pipes to all rooms. Tape, wrap, and install register boxes on all lines. Refer to the heating layout from your furnace supplier to install the proper-size duct to the proper room.

ROUGH-IN PLUMBING Your home is framed, with all walls in place and plywood on the roof. If you have a concrete floor, you now connect to the ½″ copper water lines extending up out of the concrete floor at each fixture location. Now, rough in or install the necessary addition-pipe in the wall, extending out through the wall for fixture connection later.

SEEPAGE PIT If your home must have a septic tank, your overflow drain from the tank can be a seepage pit, or if the code allows, a leach line. Seepage pits are sometimes needed on smaller lots that do not have space for a leach line. A seepage pit is a factory-made concrete tile 5′ in diameter and usually 4′ tall. These are stacked on end, one on top of the other, down into the ground, sometimes for 20 feet, depending on the type of soil, to achieve proper drainage. A leach line is most economical and the best choice.

SEPTIC TANK If you do not have a public sewer available from your property, you will need a septic tank, a leach line and/or a seepage pit. In most areas, a two-bedroom home requires a 1,000-gallon tank. A three-bedroom home takes a 1,200-gallon tank. Septic tanks are available in concrete or, in many areas, fiberglass. Both are good.

SETBACK LINE In most areas, your home must be a certain distance back from the street and from your side-yard property lines. You must determine your local restrictions before building.

SHEATHING Sheathing normally refers to roof plywood. However, if your home will be brick veneer or exterior siding, and you live in a cold climate, you may be installing Celotex sheathing directly to the studs on all outside walls.

SHUTOFF VALVE This valve is used where the water supply from the street connects to the outside of your home. In case of plumbing problems, all water can be shut off from the home. A shutoff valve must also be installed at all fixtures on both hot and cold supply lines.

SIDEWALKS AND DRIVEWAY Front-entry sidewalks are usually 36″ wide, 4″ thick and of poured concrete. Rear-door or other walks can be 24″ to 30″ wide, also of concrete. Driveway width is determined by the width of the garage door you have installed. On both walks and drive, use 2 × 4's on edge for forming. After the concrete has been troweled properly, a stiff broom can be pulled across the slab to produce a broom finish which eliminates slickness when the slab is wet.

SIDING Siding is the finish material on the outside of your home. This siding comes in many forms: sheets, lap, horizontal, etc. Under this siding, especially in colder climates, should be installed the sheathing of Celotex, plywood, etc.

STUDS This framing member is usually a 2 × 4, precut 92¼″-long at the mill. It is added to the thickness of the bottom wall plate and the double wall top plate to figure the 8′ ceiling height.

STYROFOAM INSULATION   This insulation is normally required by code in cold climates, and when there is a concrete slab floor in the living area. Refer to foundation details covering these conditions. Some areas require this 1″ Styrofoam panel to be glued with mastic only to the outside of the footings. Other areas require Styrofoam both inside and outside the footings, as shown on the details.

SUMP PUMP   Basement homes should have a sump pump to discharge water from the basement in case of flooding. A 16″-diameter concrete tile should be installed on end, down through the basement floor, as a catch basin for water. Install a standard sump pump in this tile and connect a discharge drain to the outside.

TEXTURED WALLS   After drywall has been installed on all the inside walls of a home, all joints must be taped and filled. Also, all nail heads must be recessed and filled with joint mud. All joints and nails must then be sanded smooth before wall texture is applied. With special equipment and an air compressor, a subcontractor will spray all walls to an orange-peel texture, ready for painting when dry. Walls will be textured after the acoustical ceiling is applied.

TOP PLATES   On top of every wall, install a double 2×4 top plate. This extra 2×4 is required to help support the roof load resting on the walls.

VAPOR BARRIER   Some areas require Visqueen (plastic sheeting) under a concrete-slab-floor home. On a crawl space type home, this Visqueen is also required, lying on the earth and covered with 2″ of pea gravel. On basement homes, Visqueen is installed under the floor slab and glued to the outside basement walls before backfilling is done.

VENTS (Attic)   In each gable end of the home, you will normally install a 15″×24″ screened vent to allow air to flow through the attic area. In the roof overhang, you can install 8″×16″ screened vents every 8′ around the home. In some areas, you will use a vented blocking between rafters.

VENTS (Crawl Space)   In the top row of blocks in the crawl-space wall, install an 8″×16″ screened and louvered vent every 8′ around the home and at the corners. This allows ventilation under the wood floor.

VENTS (Plumbing)   Projecting through the roof approximately 12″, you will have a vent pipe with the roof jack. This pipe is usually ABS plastic pipe. Every fixture must be vented to the outside.

VERIFY ALL DIMENSIONS AND CONDITIONS ON JOB   Throughout the country, there is a Uniform Building Code. This book is used in most cities and counties. However, because of variations in climate, foundation codes vary from state to state. Other additions, code changes, etc., have been made to suit the conditions in specific localities. The user of this book is responsible for learning about and conforming to any and all building codes or requirements of the city, county, or state in which his home is being constructed. He must verify with his local building department all dimensions and conditions on the job site.

WALL CHANNEL   This wall-framing member or channel is built from three full-length 2×4 studs. These studs are nailed into a "U" shape as shown on the drawings (page 73). This assembly is nailed into a wall at the exact location where another wall, at a right angle, joins the first wall to form a corner of a room. The bottom of this "U" is always turned to fit against the end of the new right-angle wall. This channel helps to plumb the right-angle wall.

WINDOW WELL   One of these is used on a crawl-space home. At the access door, which is usually a standard basement window, install a window well for access to the crawl space. On a basement home, install a well at each window. This unit reduces the height from the finish wood floor down to the finish grade, giving the home a more pleasant appearance from the street. It allows the basement window to be recessed below grade, out of sight, yet supplies light and ventilation to the basement. Window wells are formed from sheet steel and have a galvanized finish to resist rust.

WIRE MESH   Many parts of the country require this wire mesh to be installed in concrete floor slabs, garage floors, and patio slabs. It helps eliminate future cracks in large slabs, due to weather changes. This mesh is made by factory welding of ³⁄₁₆″ steel rods into a pattern of 6″ square openings, and is normally fabricated in 6′-wide rolls 100 ft. long. To install this mesh, unroll it and lay it flat over the entire area to be covered by the slab. Next, raise the mesh 2″ above ground by placing blocks under it. This will allow the concrete, when poured, to wrap around each steel bar.

"X" BRACING   If your home will have a wood floor with floor joists, you must install "X" bracing between all floor joists. This can be made from 1×4, 2×4, or metal "X" bracing, which can be purchased locally. This bracing is installed from the top; side of joist to the bottom; side of the joist over. Then, from top back over to bottom, side of first joist. This forms an X between all floor joists. See Section View Through Basement, page 48.